Eric Orff, CWB

# What's Wild

A Half Century of Wisdom

From the Woods and Rivers

of New England

*Peter E. Randall Publisher*
*Portsmouth, NH / 2024*

© 2024 Eric Orff
All Rights Reserved.

ISBN: 978-1-942155-76-8
LCCN: 2024911887

Published by
Peter E. Randall Publisher
5 Greenleaf Woods Drive, #102
Portsmouth, NH 03801
www.PERPublisher.com

Cover and book design: Mindy Basinger Hill

Printed in the United States of America

I am forever grateful for the help and support

of my family over the years, but especially Janice Orff,

who has been my life partner and my wife for over a half century.

She has supported my work over the years and urged me

to publish this book.

Thanks Jan.

# Contents

Preface / ix

Hi-Ho Whitetail . . . Away / 1

I Stopped a Fish Kill Today! / 8

New Hampshire SSSSnake SSSStories / 13

The Tale of the Mysterious Duck Boxes / 17

The Day the Suncook River Flowed Upstream During the Great Mother's Day Flood of 2006! / 20

"Hello, There is a Moose in My Swimming Pool" / 24

And Suddenly It's May! / 27

My Tree Stand, My Castle / 31

The Slippery Slimy Scary American Eel! / 34

Me and Three Bears Up a Tree / 37

May I Have This Dance? A Waltz With Fire / 41

Diving Into the World of the Dead / 44

What is a Hunter? / 48

Henry . . . Just Henry / 52

With a Cluck-Cluck Here, and a Gobble-Gobble There / 56

New Hampshire Conservation Legends / 59

Catching the Spirit of the Native Brook Trout / 61

The Long Journey to the Opening Day of Trout Season / 64

Ivory Floats / 66

Weird Woods Happenings / 69

New Hampshire's "Rodney Dangerfield" Animal:
The Fisher / 72

How Much Wood Can a Wood Chuck Chuck??? / 74

New Hampshire's Mid-Winter Coastal Waterfowl Survey
Takes Flight / 77

I've Got Friends in Low Places / 82

How Old Is Old in Animals? / 85

Fish and Wildlife Restoration in New Hampshire:
A Century of Successes! / 87

Sowing the Seeds of Conservation That Will Last
a Hundred Years / 91

Fish Climbing New Hampshire's Ladders of Success / 96

Fishermen Catch Tons of Tasty Treats on New Hampshire's
Tidal Waters / 99

What's Up with the Clams Down on New Hampshire's
Seacoast? / 101

Global Warming Threatens New Hampshire Hunting
and Fishing / 103

If Only Moose Could Vote / 106

Downwind and Dirty: Living in the Shadow of a Coal-fired
Power Plant Stack / 109

Clean Water and Air is a Must for Our Fish and Wildlife / 112

Do Bears Hoot? Bear Myths and Medicine / 115

Go Take a Hike . . . at Night / 117

New Hampshire Stripers Are for the Birds! / *120*

Guess Who's Coming to Dinner? / *123*

New Hampshire Wildlife Flourishing / *127*

22 Million Juvenile River Herring Seeking the Sea / *133*

New Hampshire's Threatened and Endangered Wildlife Rebounding from the Mountains to the Sea / *135*

Heavenly Habitat / *137*

Silent Woods / *139*

Bear-Paw Greenways: Woodland Tracts for Tracks / *141*

It's Duck Season in New Hampshire / *143*

We Have a Lot to Be Thankful For in New Hampshire / *146*

Endangered and Threatened Birds Making a Comeback in New Hampshire! / *149*

The Eye in the Northern New Hampshire Sky / *151*

RX-deer: Taking the Pulse of New Hampshire's Deer Herd / *153*

Wild Goose Tails / *155*

Just Ducky in New Hampshire / *159*

2022 Was a Banner Year for New Hampshire Hunters / *162*

About the Author / *164*

# Preface

I have been a conservation communicator for my entire half-century career. Many of the articles in this book offer a historical perspective of my three decades as a wildlife biologist for the New Hampshire Fish and Game Department, as well as my fifteen years working for the National Wildlife Federation (NWF). The purpose of the NWF work was to educate and motivate others to act, as the impacts of climate change on New Hampshire's fish and wildlife became more evident to me.

When I first began writing and sharing my observations as an NWF field biologist, my audience of hunters and fishermen were reluctant to hear the message, despite the credibility that I had earned after writing accurate information for decades. I knew it was important to be clear and share accurately what the science was telling me, and over the past twenty years, the acceptance gained traction.

I began working at the New Hampshire Fish and Game Department in 1976 as a fisheries technician, subsequently becoming a fisheries biologist before getting the position as the department's first black bear biologist in October 1978, a position I held for over two decades. I became the department's furbearer biologist in 1980, a position I held until retirement in 2007.

From 1983 to 2005 I owned and operated Bat and Wildlife Control Specialists, a nuisance wildlife control business. I specialized in nonlethal control of bats, but also worked with lots more wildlife at this part-time business. From mice to moose, I have handled them all.

Shortly after retirement from the New Hampshire Fish and Game Department I was recruited to work as a consultant for the NWF as a field biologist, a position I held for fifteen years.

I have been a member of the New Hampshire Wildlife Federation since 1964, starting at age fourteen, serving as a director for numerous decades. I received the New Hampshire Wildlife Federation's Sportsman of the Year award in 1998.

During the last decade of my career with the New Hampshire Fish and

Game Department, I wrote and voiced public service announcements about fish, wildlife, and environmental topics, producing three timely and pertinent topics per month in cooperation with WOKQ/WPKQ radio station. This resulted in in-kind donations of over one million dollars to the New Hampshire Fish and Game Department. As a result, my voice is often recognized today, even at the grocery store!

I became a member of the New England Outdoor Writers Association, and then served on the board of directors from 1998 until 2018. I received their Dick Cronin award in 2004, among other writing awards and commendations.

Since my retirement I have expanded my educational outreach through my social media networks and as vice president and spokesperson for the New Hampshire Wildlife Federation (NHWF). My multiple weekly posts focus on my timely observations to inform my followers of the rhythm of nature in New Hampshire.

During the pandemic, I took it upon myself to develop the social media presence of the NHWF. The effort has been successful—particularly with Facebook. We have grown the number of followers from about one thousand in 2020 to over twelve thousand in 2024. The NHWF Facebook page is averaging over a hundred thousand views a month, with some months hitting over a million views.

I enjoy supporting various groups and I have held board member positions for over ten years each at both the Bear-Paw Regional Greenway and the Friends of the Suncook River. In my home community, I was a member of the Epsom Conservation Committee for over a decade.

I reach out to my followers nearly every day to provide information and observations, while striving to engage New Hampshire citizens to take action on climate change issues. I serve on the *Concord Monitor* newspaper's reader advisory board to advocate for conservation education. I am currently on the New Hampshire Audubon policy committee. My goal is to also motivate our elected representatives to take action.

The information in these articles is accurate for the time they were written. I have provided some updates as comments to many of the articles to bring readers up to date as of the publication in 2024.

# Hi-Ho Whitetail . . . Away

In my nearly thirty years as a wildlife biologist for the New Hampshire Fish and Game Department I have had many wonderfully exciting experiences dealing with all sorts of wildlife. Since 1983, I have also owned and operated a part-time nuisance wildlife control business that provided another whole spectrum of problems. Certainly, a thousand or more of these have brought me up close and personal with bear, moose, deer, bats, snakes, bobcats, and more. I have literally had experiences with animals from mice to moose. Fortunately for all of that time I have kept a daily diary and have recorded much of those encounters in detail. Winter is a good time to turn back the pages of time to relive some of those moments.

Some experiences stand out far and above most others and for me a couple of cold November weeks in 1983 is one of those. You see, I was picked by my supervisor, Henry Laramie, to help him capture and remove live deer from Long Island in Lake Winnipesaukee. Long Island is about a half mile wide by one and a half miles long with a deer population that exceeded a hundred at that time.

To give you a little background, Long Island is an island that lies just offshore from Moultonborough Neck, connected to the mainland by a bridge. For many years it held the densest deer numbers in the state. In fact, surveys done at least twice had counted over one hundred deer. I had participated in a couple of "island counts" on the islands, where UNH students were bused up and formed lines to drive the deer past counters to tally them. I had participated in these counts as a UNH student on Long Island as well as on Big Diamond Island on the same lake in the early 1970s.

Plus, for years I had been told stories of the famous Long Island special "paraplegic hunts" by a friend and mentor, Bill Boucher, a disabled Korean War veteran from Londonderry, the town where I had grown up. Bill was in a wheelchair and had participated (to say the least) in these hunts through much of the 1960s. These hunts helped to keep the deer herd in check for a while but had stopped by 1970. Since the hunts had been stopped, the deer herd had increased to over a hundred and were stripping the island of its vegetation. In fact, I stayed in Bill's camp on the island for a night in 1970 or 1971 and hunted the mainland nearby hoping that some of the island deer had wandered there.

However, even though a few deer did cross back and forth, most stayed on the island and their numbers had pretty much cleared the island of its understory vegetation. Deer were starving in late fall when they should have been plump for the upcoming winter. All I know is suddenly a political decision was made, not to hunt the deer but to have Fish and Game staff remove some. Lucky for me, Henry picked me to go, so I hurriedly got my stuff together and began collecting the equipment needed to tranquilize deer. I did have three or four years' experience tranquilizing nuisance bears with Henry, so I had a leg up on most other biologists within the department.

## Monday, November 7, 1983, clear nice day

My diary says:

> "I finished collecting supplies at the Fish and Game headquarters in the morning and headed up to Long Island by 1:00 p.m. We set a corral trap made with a huge net borrowed from the Fisheries Division and also had 8 big box traps, called clover traps, made from tubing and netting that were specifically designed to catch deer. We baited the traps with apples and waited for darkness when the deer would be most active. That first day we missed several deer but did catch a spike buck, which died."

These deer were actually in terrible physical condition from the starvation diet they were on because there were way too many deer for the habitat to support. That's why we were there, to reduce the numbers by capturing the deer and relocating them. Even by giving them the best of care we had a number of deaths. They were simply too malnourished to be captured or tranquilized. At least a quarter of the twenty-seven deer we captured died before they could be released. The ones we caught in the nets we quickly put into a transportation crate without tranquilizing them to reduce stress. They were shipped south within hours. Even several of these deer died.

### Tuesday, November 8, clear cold night

> "I had come home after the captures Monday night but headed up mid-morning today with more warm clothes and items needed for a long stay. The deer started moving about 3:00 p.m. Henry caught a doe in a drop net and we missed one in the corral when it bolted just as the gate fell. Later we successfully captured another doe in the corral. The director Charlie Barry stopped by tonight and two conservation officers, trainee Tim Acerno and Lieutenant Dave Hewitt, stopped by to assist for a while."

Over the next couple of days we managed to catch two or three deer in various traps each day. In fact, I had brought up some radio collars and telemetry with me and I rigged the collars to begin transmitting a signal if a trap was set off. That way we could run or drive quickly to the traps to prevent escapes.

We had brought an old Fish and Game camper trailer to sleep in. Henry got the bed and I got the floor in this peanut-sized rundown tin can. I was so excited trying to catch deer that I just couldn't sleep anyway. A couple of times the receiver started pinging at 3 or 4 a.m. and I darted out of the trailer to find a raccoon in one particular clover trap.

Friday, November 11, is one of those days that stand out in my mind. Not for what we caught, but for what we didn't. Henry had left for home to get more supplies around 1 p.m. Tim Acerno helped me move a couple of clover traps and he too left around 2 p.m.

> "By now the deer seemed to have learned about our traps and were staying clear of them. I lay down early last night and was awakened about 9:00 p.m. by my radio system. It was the "raccoon trap" pinging.

*I headed out alone expecting to release yet another raccoon. Not!! A huge buck, ten points by my up-close count, was flailing in the trap. Boy was he unhappy to see me! He was lifting the trap into the air and I was worried he was going to run off with the trap. So, all alone in the dark I tried to hold it down as his antlers thrust in every direction, including mine. I tried to get a hand in my pocket to get out a syringe and drug. Things were not going well at all. By now all the tie downs had come off the trap and he was getting more rambunctious by the minute. He and I were in a standoff for a brief moment, then he simply put his head down, ripped the side off the trap, and disappeared into the bleakness of the night."*

I knew there were some big bucks around as I had used a Vietnam-era night scope to watch a ten- and twelve-pointer duke it out in a field not far from here. In fact, I had watched bucks fight several times while watching the clover traps to see if a deer went in one, before I came up with the radio collar idea.

## Tuesday, November 15, rain to snow at night

The director had ordered us to begin shooting deer with the tranquilizer gun as they were now pretty much all trap-shy. This night had me trying to catch deer that we had hit with a dart but hadn't gone down. I had started checking traps at 6 a.m. and ended up working practically all night as well. My diary says 22.5 hours straight. And I loved every minute of it!

Within a day or two of losing several darted deer, our order of "radio" darts arrived. These were simply miniature radios that we screwed into the base of our darts that sent out an electronic signal similar to the radio collars. Now we had the equipment we most needed. Since I had the experience tracking animals with radio telemetry, it was always my job to find and hog-tie the deer that were darted with the radio darts.

Let me tell you a little bit about the drug we were using to tranquilize these deer. It was and still is one of the best available, but under field conditions, lots of things can, and did, go wrong. First of all, even when tranquilized the deer were sensitive to light and were especially sensitive to any sharp noise. Consequently, when I headed after the deer, I would turn my flashlight off and not turn it on until I had the deer tied up. I could tell from the signal generally how close I was to the deer. When I got close, I would crawl on my hands and knees in the dark to feel for sticks so I wouldn't make any sharp sounds. Time after time I put my hands on deer in total darkness. I learned rather quickly to first get a rope around the deer's neck and tie it to a tree so it couldn't escape.

I remember one night putting a rope around a deer's neck, just then, and before I could tie it to a tree, it ran off trailing my rope. How I listened as the deer dashed off into the night. I spent the better part of the next hour in a cold cloudy moonless night quietly crawling and feeling the forest floor until I found the quarter-inch rope. Those were adrenaline-filled nights!

## Thursday, November 17, cloudy but cleared by night

This was to be one of the longest and most memorable days and nights of my career. We had darted a doe on the evening of the 16th and lost it. Everyone else had gone home but I stayed alone and couldn't sleep thinking about the lost deer. By 10 p.m. on the 16th I was back out searching for it. The deer was slowly staggering but managed to stay just ahead of me. I was tracking it with the radio but it wouldn't lie down long enough for me to grab it. I remember following the deer just a few feet behind it as it crossed someone's lawn. I had to bend over when I walked past a picture window within arm's reach of the TV the folks inside were watching. Well, I finally did catch up with that deer around three in the morning. In the process of catching and hog-tying the deer, I managed to lose the new Fish and Game portable radio out of my pocket. When conservation officer Chuck Kenney arrived about seven the next morning to check on me, I told him, "I've got some good news and some bad news. I caught the deer but lost the director's new portable radio." Boy, was he concerned about the lost radio. Somehow, I found the spot in the middle of the woods where I fought the deer the night before and found the radio buried in the leaves.

By nightfall we were cruising the roads trying to dart deer from the vehicles using our usual technique of spotlighting. I was with Henry. He had loaned his tranquilizer gun to District Chief Pete Lyons and Conservation Officer Chuck Kenney. The radio crackled that they had darted a deer and wanted me to come find it and tie it up. Our general rule, while darting, had been to only dart does since they are the ones most likely to add to the island's population.

About 9 p.m. I arrived to meet them. I soon set out into the darkness to find and hog-tie the deer. In short order I found him. Yes, *him*. Even in the darkness my hands told me this was a very big buck. I got my rope out of my pocket and tied it around the deer's neck and tried to tie it off to a nearby tree. No luck, as no tree was close enough. I figured I'd just sit on him and signal for help. I did by flashing my light. Since there were three others, I figured they could help hog-tie him. Just as Chuck got near, he stepped on a stick, sending a *crack* into the cold night air.

Suddenly, I mean very suddenly, the buck was headed downhill with me on his back! I remember gripping his antlers very tightly. All at once I was transported into some Star Trek episode as I found myself at warp-speed with trees, boulders, and indeed my very life flashing by.

Just as suddenly, he stopped. I didn't. But I didn't let go! Now I had this huge buck in a very firm antler-lock as he was trying to shred me with them. I almost threw him to the ground once, but he bounced right back up. Finally, I ended up against the hillside holding on to the antlers as he tried to drive them into my crotch.

By now I could tell this buck was getting pretty mad, as he clearly was trying to send me to the soprano section of the choir! I wasn't about to let him go so he could build some momentum. It was only a few nights before that I had witnessed a similar-sized buck run off into the night with the remains of a clover trap hanging from his antlers. I surely didn't want to see this buck running off into the darkness trailing parts of my anatomy that I considered very valuable.

I yelled to Chuck, "Take out his back legs, take out his back legs." Chuck lunged behind the buck, but before he could catch a leg, the deer kicked, sending Chuck flying backwards down the hill and out of sight. However, soon the other two arrived and we managed to bowl the buck over. I had him tied up pretty fast.

This ten-point buck bottomed out the two-hundred-pound scale when we tried to weigh him while ear tagging him. Pete said, "I just wanted to prove to Henry that there were some big deer out here too." Indeed there was and my crotch had been transformed into eight black and blue widely spread points to prove it for over a week afterwards. I was very sore for a few days. We wrapped up the deer removal the Sunday after Thanksgiving, the 27th. In all we had removed twenty-seven deer from the island.

This was one of the most memorable times of my career and I savor reading my diary from time to time to relive every deer captured. I still have the antlers we sawed off that buck hanging in my garage. In fact, I once rattled-in a huge buck with those antlers, but my muzzle loader misfired. Now that's another whole entry in my diary...

*First published in 1983.*

# I Stopped a Fish Kill Today!

Tuesday, October 16, 2001, 8:45 a.m., Suncook, New Hampshire. I arrived at the lower dam on the Suncook River to check on the downstream migration of literally millions of river herring or alewives. There has been an ongoing effort for nearly a decade to restore alewives to the Merrimack River, and the Suncook River has played a pivotal role in this restoration effort.

Historically, alewives swam up the Merrimack River and its major tributaries by the untold millions each spring to spawn in the quiet waters of the lakes forming its headwaters. After living in the sea for two to three years, and growing to ten to twelve inches long, the adult alewives make a mad dash up the rivers to lay as many as two hundred thousand eggs per female. Native Americans caught multitudes of alewives with nets all along the Merrimack River system. They also constructed special fish traps, called weirs, that were set along the rivers. Weirs Beach on Lake Winnipesaukee retains in name only the significance of a time lost. The runs of alewives were lost as well, beginning with the construction of dams along the tributaries. On the Suncook River, in what was to become Allenstown and Pembroke, and locally known as Suncook, the first dam was constructed in the early 1700s. The waterpower served a local grist and sawmill. In 1848, a huge dam was constructed across the Merrimack River in Lawrence, Massachusetts, seemingly forever altering the upstream migration of fish. Within two or three years the tremendous annual migration of Atlantic salmon, American shad, and alewives were lost.

However, all was not lost, for efforts have been underway for over three decades to restore shad and salmon to the Merrimack. The salmon restoration has been met with limited success; however, the shad restoration has been very successful. Over seventy-six thousand shad were passed over the fish passage facility at the dam in Lawrence in the spring of 2001.

Alewife restoration efforts had taken a back seat to the more noteworthy game species, salmon and shad. A restoration effort was finally begun in 1994. New Hampshire Fish and Game fisheries ecologist Bill Ingham initiated the effort by transporting two hundred adult alewives from the Cocheco River in Dover to Northwood Lake in Northwood. According to my diary the fish were released about 2 p.m. on April 24, 1994, as I accompanied Bill. I have had a lifelong interest in the Merrimack River, since I grew up in Londonderry and often fished the local brooks for trout until they entered what was then the open sewer of the Merrimack. I have lived on a bluff overlooking the Suncook River in Epsom since 1979 and have taken a keen interest in it as well. Although alewives are not considered a game fish, they provide a tremendous forage base for nearly all the other larger fish. From the lakes where they are spawned and grow down through the river system as they migrate, and especially out in the sea, alewives are providing nourishment to others. Each year some of the most productive fishing for striped bass occurs in the lower Merrimack as these giants follow the migrating alewives into the river. I watched three river otters indulge themselves in a school of juvenile alewives above the Northwood Lake dam on October 2. The return of alewives would help bring the entire system back into balance.

Since 1996, the US Fish and Wildlife Service fisheries staff have stepped up the transfer of fish from the coast, stocking several thousand adult alewives into the two major tributaries of the Suncook River, Northwood Lake and Suncook Lake in Barnstead. Thanks to this effort, several million eggs hatch each spring in these lakes and by September millions of juvenile alewives are nervously awaiting their trip down the river and out to sea. The hordes of fish (huge schools often can be seen by late September at the dams) are released to migrate downstream when the dam boards are removed each fall to lower the lake levels for the winter.

The dam boards were lowered at Northwood Lake on October 6. Because of the huge numbers of fish that have raced out of the lake in past years when the dam boards were pulled, this year the public was invited to watch. A newly formed conservation group, the Friends of the Suncook River, hosted the "Million Minnow March" that rainy Saturday in anticipation of the spectacle. About seventy people crowded around the dam to stare at the torrent of water leaving the lake once the boards were removed. But fish being fish, they swam their merry way and elected not to join the festival. Residents who lived on the lake and came to watch, have marveled all summer long at the huge schools of minnows swarming along the shore.

The migration began four days later on October 10. By Thursday the 11th, a few speedy alewives were well down the river and were passing the two hydro stations on the lower Suncook at Pembroke hydro and the China Mill hydro station. Because of the low flows in the river, these plants had not been operating. It is the lowering of the lakes above that sends a torrent of water and juvenile alewives downriver and also enables the hydro stations to begin to generate electric power: sometimes a deadly combination.

By Friday afternoon the migration was well underway. I enjoyed an unusually warm mid-October evening on the banks of the Suncook at a cookout at a neighbor's picnic spot at the river's edge. As far as I could see up and down the river, for over a hundred yards in either direction, the characteristic dimpling of the surface by juvenile alewives could be seen. We all marveled at the sight for the remaining hour or two before dark. Even then, when I took a strong flashlight and cast its beam onto the river, I witnessed the stampede continue into the night.

The next afternoon I stopped by the China Mill dam to watch the fish tumble over a bypass gate near the intake of the hydro station. The turbines remained still and hundreds upon hundreds of little silver streaks were swept over the spillway and were quickly carried down and deposited not one hundred yards from the Merrimack River. I watched the race downstream for nearly an hour. At one point I counted about three hundred fish a minute going over the falls. Most of the water was still spilling over the main dam so just a fraction of the fish was sweeping by my eyes. It was still a marvelous sight to witness. After all, two thousand adult alewives were placed in Northwood Lake on May 5. Potentially five to ten million juvenile alewives were descending the river as I watched!

This brings me back to Tuesday morning, October 16, 8:45 a.m. As I approached the hydro station I could see that the operators were now generating electricity, as the current was rushing into the power canal that delivers the water to the huge intakes to the turbines, now whirling away. But my heart sank as I noticed that the bypass was not opened *and* right in front of the intake grill two large schools of juvenile alewives were swarming. I quickly made my way around the canal and a second glance at the water showed it devoid of fish. They had been swept into the intake's powerful grasp.

I quickly made my way around the back of the huge brick factory that has stood here for over one hundred and fifty years, drawing its strength from the river itself. It is no easy task to get to the discharge end of the hydro station. You must fight through a jungle of growth, slip through and under a portion of the fence where the river has eaten its foundation, and then traverse the narrow corridor separating the factory and the river. Even this stretch is a challenge. I fought my way along the banking and sprinted past the two pipes three stories over my head spewing steam and fabric debris out their snouts like angry dragons. It was obvious that sometimes a great volume of scalding steam and debris is discharged onto the path I had selected.

At the tailrace of the hydro I saw what I expected to see . . . fish dying by the hundreds. Some fluttered in a fruitless and vain attempt to swim with their torn bodies in the swift current. The bottom of the discharge canal was littered with the silver glimmering bodies of those that had succumbed to the turbines minutes before. Most fluttered by in a steady stream holding on to their last minutes of life. Some were just bits and pieces being swept along and tumbling in the current at the bottom of the canal. I saw no survivors.

This was the same bleak image that I had seen before, in 1996. That year, while on the opposite side of the Merrimack River from where the Suncook River enters, I noticed a huge swarm of gulls diving and feeding. I drove over for a closer look and discovered a massive fish kill. Perhaps millions, as the whole bottom of the Suncook River and well out into the Merrimack were littered with dead alewives. There were no gulls here, so very likely this fish kill had just started. There was far less death this time, but clearly thousands of once vibrant little silver-sided baby alewives were strewn down the river in a scene of mass carnage.

I quickly made my way back around the old mill building and was fortunate to meet the hydro operator as I arrived back at the intake. "You're chopping up the fish," I exclaimed. He quickly shouted to another fellow to run and shut it down. At 9:02 the intake pipe belched back its last gasp of leaves and young fish. The fish kill was over!

The hydro operators at this site and the one upriver have agreed to bypass as many fish as possible. Usually, water is allowed to go past the intakes and many of the fish can continue downstream unharmed. This operator said that he had been watching for fish and regularly checking the canal. However, a school of fish did manage to slide down to the intake unnoticed. Both this and the next upstream hydro facility, the Pembroke Hydro Station, are under new ownership and they feel very strongly that all the fish should be allowed downstream unharmed. Fish and Wildlife Service biologists have been working with them to minimize fish losses by recommending flow bypasses and other diversion plans. The hydro operators, to their credit, have greatly reduced their generator capacity by slowing down the turbines so the fish are not sucked into the intakes and then the fish can be directed over the dams by allowing a significant amount of water to bypass the intake and carry the fish away.

Progress is being made but there is much that remains to be done to restore the Suncook and Merrimack Rivers. The Friends of the Suncook River hope to be a part of a solution to fish losses that are plaguing the restoration efforts of alewives to the Merrimack River system.

*As of 2024, the efforts to restore salmon to the Merrimack River have failed and been abandoned. Efforts to restore the shad continue.*

*First published in October 2001.*

# New Hampshire SSSSnake SSSStories

Since I can first remember, I have had an interest in all sorts of wildlife, and snakes have always held a special fascination. The first snake I can vividly recall was too quick for me as it slithered then swam away from me as I chased it in a creek in Oklahoma, when I was four years old. Good thing. It may have been poisonous! Luckily for me, as well, has been the fact that I grew up in Maine and New Hampshire, where poisonous snakes are practically nonexistent.

While I happen to like snakes, I have met lots of people who didn't. Some of these folks were actually terrified of them. Let me tell you about a few of these encounters. Since 1983, I have owned and operated a small part-time nuisance wildlife control business called Bat and Wildlife Control Specialist out of my home in Epsom. Although I primarily deal with bat problems, a few snake calls were mixed in over the last two decades. A dozen or so snake encounters I have had come to mind and some still have me chortling. I've just got to tell you my two favorites right up front.

I took a call one cool rainy spring morning from a fellow in Hopkinton. I usually tried not to work the weekends as by then I had already put in a full week at my full-time job as a wildlife biologist with New Hampshire Fish and Game, plus my part-time work. But this fellow tugged at my sorrow side as he nearly wept on the phone. His wife had run off. And I'm pretty sure with her pants down! He was desperate to get her back. But a snake stood, or should I say, hung, in the way.

His beloved wife had entered the bathroom that morning to do what we all need to do, and was gracefully seated on the throne, when she happened to look up. Her shrieks and a parting blur as she streaked out of the house was his last sign of her. She did mention she was not coming back until "it" was gone. Now *it* just happened to be a very harmless eighteen-inch-long milk snake. But it was *it's* penchant to lie on the suspended grill of the recessed light directly above the toilet. That's what led to his wife's sudden urge to streak. Rather late in life I must say. The glowing warm light fixture served as a perfect sunning site for the cold-blooded serpent. *It* had likely wintered in the stone walls of this old colonial house.

Although the husband had already tried a capture by the time he called me, the snake had slithered back into the ceiling. I told him to leave the light on and I would be right over. An hour or two later I arrived and carefully sneaked into the bathroom. The lush snowwhite carpet cushioned my steps as I recall. Sure enough, *it* was back sunning itself. I grabbed the snake gently with my forceps just as *it* tried to make an escape. I actually had to dismantle the light fixture a bit to untangle the writhing mass.

I checked around the foundation and sealed what cracks in the old foundation I could find. Milk snakes, like mice, tend to gravitate to these older stone foundation homes to winter. In fact, I think it is the abundance of mice that attracts the snakes.

Story number two was centered in a similar vintage Cape-style home in Lee. It too had a stone foundation. In fact, I ended up at this house twice, but years apart. The first time I pointed out to the owner the cracks in the foundation and snake skins and mouse droppings in the attic and recommended some fixes. He assured me he could/would do them himself.

*Not.* So, five or six years later I found myself pulling into a familiar driveway. The new owner had pinpointed the "nest" of snakes and wanted me to grab them and run. The nest happened to be right above the exposed beams of the living room where snakes had a habit of traversing the beams as those below were watching TV. A real *Wild Kingdom* show without Big Jim to wrestle them. I was to be Big Jim. The owner had his skill saw all connected to a power source and demanded that I should grab the snakes once he cut the wide old boards of the attic floor to expose them. Surprise, there were still milk snake skins and mouse droppings in the attic just like the last time I was there a few years before.

Now, no snake is going to lie in wait as someone cuts through boards with a powered saw. I was certain of that. But from the intensity of his efforts and the look on his face I knew better than to offer any advice at this point. Sure enough, no snakes were coiled ready for my capture when he pried up the flooring.

"But wait," he exclaimed, "the nest must be behind the kitchen cabinets!" He was so anxious to rid his house of the snakes that he didn't want to waste time removing the dishes from the cabinets. He and I simply grabbed and tore the cabinets from the walls as the dishes ricocheted inside. Nope, we found no snakes there either. We looked high and low. From my rough estimate we did several thousands of dollars' worth of damage to his house that evening. I kept mentioning the need to get rid of the mice and to seal the foundation. He never listened either. I bet there are still milk snakes in that house. But one thing is for sure. He's not!

Then there were the snakes popping up into the apartments of the lower floor of a large apartment complex in Derry. Just so happens there was a drainage system under the floor of these below-ground units. It also happened to connect to a drain at the bottom of the outside staircase leading down to the lower level. I determined that garter snakes were migrating from the steep hillside in back of the building and were being funneled by the parking lot curbing into the stairwell. During their fall migration they were popping up like magic from the drainpipes all over the lower floor units. I simply cut a piece of quarter-inch hardware cloth and covered the outside drain to prevent them from getting into the pipes.

That's not all of the snakes that have popped out of a pipe. In Concord, I went to the home of an elderly couple where snakes were coming out of the toilet. That's right, I'm sure glad the lady in the first story didn't live here. But it's true, small garter snakes were appearing in their toilet like clockwork. As it turned out this couple had decided not to hook up to the city sewer line when it was constructed in front of their house, due to the cost. But their septic tank had been dug up in their backyard for some time. It was a six-foot round hole with the old rusty cover of their tank lying across the top of the crumbling tank. It was a giant pit trap. Snakes crawled, or fell in, then couldn't climb out. So they slithered up the sewer pipe and climbed out of the toilet. This is true! I sealed the cover around the tank and propped a board into the hole so snakes, or whatever else fell in, could climb out.

Then there was the young lady who called looking for me, worried about a snake bite. She wanted me to examine her alleged bite. She had been bitten in the "upper inner thigh"! Unfortunately, my wife talked to her first and "convinced me" not to go check this poor damsel in distress. (I had a very elderly nun in a nun retirement home claim she was bitten on the breast by a bat and wanted me to check her too. No, not now, not ever!)

Speaking of nun retirement homes, I went to a snake job at one of those too. Again, snake skins, and lots of mouse droppings in an old house. Luckily no snake bite complaints there. But there was a house in Northwood where I was able to quickly open a wall and remove handfuls of garter snakes. I put them all in a bucket, brought them home and let them go. Did I mention that my wife doesn't like snakes either? Apparently she doesn't.

I did take a call once from someone who claimed they had a very poisonous pit-viper pet get loose in their house and wanted help looking for it. This was another "No, not now, not ever." But I did help the lady in Henniker who had a snake living under her refrigerator. Talk about a diet plan. I bet I could market that one. "Lose twenty pounds, this week, on my new snake-under-the-refrigerator diet. It may not all be just fat."

Being a wildlife biologist and in the nuisance wildlife control business for over two decades has brought rewards way beyond monetary. A slithering snake call is sure to bring a great story and a smile to my face.

*First published in 1985.*

# The Tale of the Mysterious Duck Boxes

The cleats I had strapped to my boots crunched the surface of the glare ice in a cadence as I strode across the marsh to check duck nesting boxes at Little Cohas Marsh in Londonderry, New Hampshire, on a gray mid-January day in 1999. This trek could have been considered routine as I had done this very same thing, about this same time of year, at this exact same location for about a dozen years.

But checking duck nesting boxes each winter, even in the same boxes, is never routine. Each box has a story to tell, a story of a life-and-death struggle, like a page from a good novel. I approach each box as if I were hurriedly reading the final few sentences of a riveting page of a novel, anxious to slide my finger into the next page and cast my eyes into another moment in time. Did anybody live here last spring? Was it a wood duck? Was she successful? How many eggs did she lay and did they all hatch? With just a little bit of imagination I can picture the little fluffy brownish black balls, which hatched only hours before, hurling themselves into the life of the marsh six feet below the entrance of the box, as the mother calls anxiously. As a wildlife biologist for the New Hampshire Fish and Game Department this is one of my annual winter tasks, to read the diary of life left the previous spring by the occupants of the duck nesting boxes scattered across marshes in the southeast part of the state. There is never anything mundane in reading the story of nature herself.

Many years, just getting to each of the nearly one hundred boxes I check can be a challenge in itself. Last year's El Niño weather brought treacherous ice across much of the state nearly all winter. Checking duck boxes meant a daily ballet as I danced across patches of good ice interspersed with open channels and thin treacherous ice. Poor ice conditions are not at all uncommon in the southeast part of the state where I work. This winter, January 1999, has brought the best ice I have seen in years, thick

*Wildlife Biologist Julie Robinson banding a female hooded merganser from a duck box.*

and white: "good ice." So I strode confidently from box to box examining the contents of each.

The Fish and Game Department has been erecting and maintaining duck nesting boxes since 1949. Back then the wood duck was just beginning its recovery from near extinction in the east. Beaver were still rare across much of the state, as was good wood duck habitat. Hollow trees were at a premium after the '38 hurricane. Wood ducks and hooded mergansers are the two most common cavity nesters using the duck nesting boxes in my area. They have rebounded to amazing numbers all over New Hampshire thanks to the earlier conservation efforts, including the placement of nest boxes. Currently our data suggests that only 2 to 3 percent of the nesting wood ducks and hooded mergansers are using "our boxes." Most reproduction occurs from the natural cavities now in abundance across the state. However, our department always recommends that foresters and landowners leave lots of old dead and dying trees, called snags, for the next generations of ducks.

Results from last winter's check of boxes shows that 448 (77 percent) of the 584 boxes checked by department staff at 96 marshes had duck use.

Wood ducks accounted for 40 percent of the activity and hooded mergansers used 55 percent of them. In northern New Hampshire the other cavity nesting duck, the common goldeneye, leave their life history in a few boxes. The successful duck use of our boxes climbed dramatically the last six or eight years as we have placed most boxes on metal stakes away from the shore reducing raids by predators such as raccoons. Raccoons have learned to search the shoreline for the delicacies contained in the boxes. Sometimes only a few feathers and crushed eggs are left from these night raiders. Not every story has a happy ending.

Not only ducks use the boxes each summer. Bird nests are quite common. The intricately woven nests of grass, reeds, moss, or feathers tell a whole other story about the boxes' occupants. Was it a tree swallow? Sometimes nuts have been cached by a squirrel or mouse. In fact, even in the dead of winter some boxes have residents. I remember one time reaching my hand in only to have a mouse run up my sleeve. I shook like a jackhammer until I untucked the back of my shirt to let the little critter escape. Some stories are a lot like real TV! A couple years ago I opened a box to find a jumping mouse in deep hibernation. He was curled into a tiny little ball and ensconced in a pile of nesting material at the bottom of the box. I gently placed him back. It was February 2, Groundhog Day! We have found many other stories in the boxes over the years. We have records of hawks, the American kestrel, using the boxes as well as evidence of owl use. Honeybees have been found too, but hornets and wasps are regulars. A couple of winters ago while I was checking boxes in Bear Brook State Park I found the remnants of a box on a metal post. A moose had walked out on to the ice and completely demolished the box. I have seen many other stories written in the snow while checking boxes. Deer, coyotes, foxes and even turkeys are regular visitors to the marshes in the winter. I even saw an albino porcupine high in an oak along the shore near some duck boxes a few years ago.

Reading the story of the duck nesting boxes will always be a rewarding part of my job. The sequels never are disappointing, as some novels can be. After all, Mother Nature has always been a great mystery writer!

*First published in 2001.*

# The Day the Suncook River Flowed Upstream During the Great Mother's Day Flood of 2006!

For much of the night and well into the day of Tuesday, May 16, 2006, the Suncook River actually flowed upstream from the Old Mill Dam in Epsom, all during the highest flood in one hundred years on the river. In fact, the river flowed north, opposite from the direction it had flowed since the Ice Age. Indeed, a half-mile section of the Suncook River, immediately upstream of the Old Mill Dam, drained during the great Mother's Day flood of 2006.

A half mile upstream, a breach in the river's historic banking was drawing the river into an entirely new channel and would literally suck the old riverbed dry in less than twenty-four hours. Not only did the new channel suck dry the riverbed above the two dams situated at the head of a series of falls that coursed either side of Bear Island, but it turned two milelong stretches of roiling flood waters on either side of Bear Island to dry riverbeds as well. In less than two days, two and a half miles of the Suncook River in Epsom simply disappeared, during a record flood!

Local dam historian Al Bickford of Epsom pegs the construction of the pair of dams, still in place at the island's north end, between 1870 and 1872. The New Hampshire Department of Environmental Services Dam Bureau names this dam the Huckins Mill Dam. Their records show it was reconstructed in 1937 for a sawmill. This dam has 202 square miles of drainage area with an impounded surface area of five acres. The western section, closest to the old mill, is ninety-eight feet wide with a maximum height of thirteen feet. Most recently, in the 1980s, the dam was converted to a hydropower unit for a short period. Then the mill site was converted into the Old Mill Restaurant and is currently occupied by the Concord Elks Club.

But let's back up to see how quickly the complexion of the Suncook River changed from a meandering river in Epsom, save for the two sections of the river that bracketed Bear Island. Instead of meandering, this section of the river tumbled down a series of rapids and falls for most of the length of the mile long island before rejoining as one at the confluence of the island's tail. This created the best section of river for trout fishing in Epsom.

The river sweeps past my house about three miles downstream. For the first time in the twenty-six years I have lived overlooking the river, it didn't completely freeze last winter, at least by my house. The snow-less winter and dry spring conditions, that just over a week ago had the sign next to the Epsom Fire Department declaring *No Burning Permits Issued*, also practically put the river into a summer slumber stage.

All that changed with a deluge of rain most of the day Saturday. By evening my rain gauge measured three and a half inches. Our sleeping neighbor was stirring by nightfall. By Sunday morning the flood warning predictions that seemed so senseless two days before were looking more ominous by the hour. The river was rising rapidly. In fact, I spent much of Mother's Day helping a neighbor downriver from me sandbag the flood wall he built in 1988 to protect the lower level of his home, which sets right at river's edge. We hurriedly added a layer of sandbags on top of his three-foot high wall. As we worked through the afternoon, on the opposite side of the eight-inch-wide wall we were capping, the river was growing in power, carrying whole trees past our trench-like view. At one point, even a whole wall of a structure with cupboards still attached bobbed by nearly eye level. The rapidly rising river rose to the top of the wall just as we finished one layer of sandbags. Then in what seemed like minutes the river won and poured through and over the sandbags. We were all soaking wet and exhausted by then and soon were resigned to the river's power over us.

By nightfall my rain gauge was full again to the four and a half inch level, bringing the total to eight inches in two days. (We got two more inches before the end of the rainstorm.) The once quiet river roared past my house in the darkness Sunday night.

*Eric Orff standing on the now dry Old Mill Dam while a 100-year flood rages in the newly made channel.*

Monday morning I awoke to a totally different looking river. It was higher than I had ever seen it; the cornfield and meadow below my house were flooded and the river was filled with debris from parts of trees to bottles and unidentified building parts. The Suncook churned all this material in a boiling rage, and seemed to spit it right at me from a hundred yards away until the ninety-degree bend yanked them from my view. It was awesome!

But not as awesome as what four locals would witness by late morning. Local farmer Bill Yeaton relayed to me the next day what he and three others, Ronnie Colby, Chris Paris, and Peter Demers, watched just as the Suncook River changed course in a matter of minutes, by noon on that Monday, May 15.

By Bill's description, the event of the century in Epsom was just underway when he arrived at what was locally called Cutter's Pit at the end of Rhodora Drive at about 11 a.m. on Monday. The Suncook had overflowed its bank a quarter mile above the access road to the pit, flowing through a wetland and boggy area along the edge of the old Suncook Railroad bed before converging on the access road and flowing into the pit. The flow was picking up dramatically by the minute according to Bill, and was soon a raging flood. The floodwaters then began to devour the bowels of the sand pit, ripping huge trees and house-sized chunks of the sand pit and quickly washing them down the newly formed river channel. The river was relentless at tearing at the earth, widening the trough to a hundred feet, then two hundred feet and more as the raging boiling flow gained momentum. Then the tongue began to devour the sand to the north along the bed of the pit slicing off three-hundred-foot-wide swaths at a time. Chomping the exposed sand pit as it marched northward within minutes, creating a canyon-like vision in the once flat sand pit bed.

It was mid-afternoon when the manager of the Old Mill site first noticed a change in the flow of water that had been pouring six feet high over the top of the dam at the Old Mill. A huge white pine tree, sixty feet in length, was forced mostly over the dam by the powerful flows on the far side of the dam. All afternoon he watched and heard the floodwaters gradually subside as it poured over the dam. By midnight only a small flow was pouring over the dam. When he last checked at 2 a.m., the top of the dam was bare.

In just over twelve hours the Suncook River breach had siphoned off all the flow, leaving the dam waterless. In fact, the river was retreating upstream flowing away from the upstream side of the dam face!

By mid-morning the water level was two feet *down* the front side of the dam—as the river continued to flow upstream. Below the dam the river was waterless; a few footlong rainbow trout were seen flapping in the few remaining pools. By day's end the channel immediately above the dam was dropping even faster as its withdrawal hastened. By Wednesday morning it was nearly possible to walk across the river above as well as below the Old Mill Dam. The growing vortex less than a half mile above the dam had swallowed all the floodwaters. The Suncook River flowed upstream from the dam to the new river channel until this portion vanished Wednesday morning. All while the flood raged on in its new channel.

*First published in 2006.*

# "Hello, There is a Moose in My Swimming Pool"

*Hello, there is a moose in my swimming pool!* This is just one of the many calls that has brought a Conflict Abatement Team (CAT) member into action. The CAT was formed in 1996 within the New Hampshire Fish and Game Department to analyze human and wildlife conflicts. Individual department members had handled many conflicts well before this team was created; in fact, nuisance bear were first captured, tranquilized and moved in the early 1960s.

Team members meet regularly to develop strategies to prevent problems and generate procedures to handle these wildlife and human conflicts which are not preventable. Wildlife conflicts are varied. In 1995, the state Animal Damage Control (ADC) agent documented twelve common animals. Bear made up 16 percent of the complaints, followed by squirrels at 12 percent, raccoons at 11 percent, skunks at 9 percent, beaver at 8 percent, woodchucks at 7 percent, deer at 6 percent, bats at 4 percent, red fox at 3 percent, coyote at 2 percent, and waterfowl at 2 percent. An additional 19 percent came from other complaints. CAT members primarily respond to bear, moose, and deer situations that might include relocation efforts.

CAT members are generally the regional wildlife biologists as well as one from headquarters and the state's ADC agent, who is the team leader: This way team members are spread across the state and can quickly respond to a problem in their region.

The first line of defense for any bear, moose, or deer problem is the Fish and Game Department's conservation officer force (COs). They are generally first on the scene and are expected to make an assessment as to whether further action is needed. A step-by-step process is helpful in determining what steps can be taken. Very often, for bear, just removing the attractant such as a bird feeder or barbecue grill is all that is needed.

Bearproof dumpsters, electric fencing, and hazing with pyrotechnics are steps that can follow. If prevention or intervention is not successful, then relocation of the animal or lethal control is considered. If an animal such as a moose is in an urban area where it poses a threat to itself or humans, then a quick relocation effort is mounted. Moving or killing the animals is generally only used as a last resort, when all else has failed.

CAT members are equipped with tranquilizer guns and restraining devices. More importantly, each member has had specialized training. In fact, this is required by federal law. Team members underwent a federally sponsored three-day training session in New Hampshire in mid-May of 1996. Capturing and handling an animal, such as a bear or moose, always puts team members and the COs in a stressful situation.

As the Region 3 Wildlife Biologist for the department, my major CAT duties have involved moose in the southeastern part of the state. Although I am the state's "bear biologist," most bear complaints are handled by the local conservation officer and the CAT team member closest to the complaint, such as the Region 1 or 2 biologist or the ADC agent. However, we all serve as backups if the closest available biologist is not available. So, in a moment's notice, I could go anywhere in the state. Such was the case this fall when I went to Stoddard for a moose stuck in the mud. Over the last two years I have been involved in ten moose incidents, including six animals successfully relocated. Seven of the eleven cities in the state occur in Region 3 as well as increasing moose numbers, so moose incidents are likely to rise.

The following is an example of a CAT situation. On July 11, 1995, a yearling cow moose had wandered into north Manchester in the early hours and apparently had, at least once, collided with a car. Commuter traffic was heavy in the area and the Manchester police were monitoring the situation. At about 7:30 a.m., the moose jumped a seven-foot-high stockade fence enclosing an inground pool and backyard on Chandler Street. She had tested the water. I was alerted at about 8:15 a.m. and arrived on the scene at 8:45.

My notes indicate that the moose was calm and in good shape in the enclosed yard, and most importantly the high fence kept the public, who had gathered in numbers, away and minimized stress to the moose. The

Manchester Police Department had done an admirable job containing the moose and keeping the public a reasonable distance away. One officer was stationed at the low point in the fence where she had gone in. He had a garden hose he sprayed at her when she headed his way; he also managed to keep her from crashing into a glass solarium along one wall of the house facing the pool. As always, the local police department plays a key role in these incidences. Keeping people from crowding the animal is probably the most important job. It keeps both the animal and the people safe.

I surveyed the fence and found a gaping hole near a back corner. As luck would have it, construction workers were building a large building next door. A simple request brought a crew over to install a sheet of plywood to secure the moose. Many thanks to them. Fish and Game Animal Damage Control Agent Rob Calvert and Conservation Officer Jim Kneeland arrived on the scene about 9:30 a.m.

We developed a plan to dart the moose safely. Again, the police department was instrumental in moving people out of harm's way in case of a ricocheting dart. Rob placed a nice shot into the rump of the moose at 9:45 a.m. Twelve to fifteen minutes later, the moose went down. Luckily it had not staggered into the pool! I quietly crept up to the moose, blindfolded it, earplugged it (a loud noise can instantly get them to their feet), and hog-tied the moose. Police officers and the Fish and Game staff carefully loaded her into the back of a pickup truck. We quickly headed up Route 93 to Hooksett. Then headed off towards Dunbarton to release her in a good piece of woods that held several other moose. No sooner did we get her out of the truck and untied that she leapt to her feet and trotted off. We all sighed in relief.

Capturing and moving an animal the size of a deer, bear, or moose is always a risky chore. Animal relocation is an option of last resort. Team effort from Fish and Game and local authorities is critical. Fish and Game conservation officers and CAT members are ready twenty-four hours a day, 365 days a year. One ear listening for the phone to ring or ready for a beep from the pager . . . *Hello, there is a bear in my . . .*

*First published in 1997.*

# And Suddenly It's May!

Thursday, May 3, 2001. 5:12 a.m. Darkness prevails as I ease myself quietly through the woods along an old railroad or trolley bed to the edge of Baboosic Brook in the town of Merrimack. I am here a half hour before sunrise to observe and record nesting activity of any ducks or geese who may have selected this area for nesting. This is one of seventy-five kilometer-square waterfowl nesting plots that will be checked by New Hampshire Fish and Game biologists this spring. For over a decade biologists have been checking these randomly selected plots to gauge waterfowl population levels. Some must be checked at dawn or dusk which brings me here at "oh dark hundred." I have been sneaking through these same trees in the darkness for over a decade. From experience I know that the next hour or two will be the highlight of the day for me.

Venus, like a gem, hangs in the eastern sky now faintly colored with a purple cast. The tongue of light now lapping the eastern horizon will soon swallow it. Peepers and toads peep and trill through the last vestiges of the night and I know they will soon be drowned out by the cacophony of bird songs that will echo through the forest. I step quietly to the edge of the brook that remains silent compared to the torrent of water rushing through last spring. I once had a beaver introduce himself to me here. He slapped his tail in the water, perhaps to startle me into moving, as I stood immobile behind a big pine tree. I could hear him sniff and sniff the air as he detected my presence, but his curiosity drew him ever closer until he climbed out of the water, not three feet from me, in the early morning light.

The quietness sends my mind reflecting about these woods, not just over the last decade, but now going back nearly forty years. My fifty-first birthday is coming up at the end of this month and I can't help but wonder how it got to be May so soon, and me fifty-one! My family moved from Maine to Londonderry in 1962 when I was twelve and over the last four decades I have come to know the woods, farms, and fields of Londonderry and the other towns surrounding Manchester.

Most would think that the last forty years would have changed the "wildness" of Manchester and the surrounding towns as the city expanded and its suburbs have seemingly swallowed much of the wilderness of the local towns. My experience has been to the contrary. Wildlife has learned to cope with humans. A statewide expansion of the deer, moose, turkey, and bear populations has spilled into the woodland and even the backyards of these towns. As the regional biologist for the southeastern section of the state since 1988, my duties have included dealing with moose, deer, and bear that have even wandered into Manchester. I have tranquilized and moved moose from the north end of the city from North Elm Street and the south from Huse Road over the last few years. The fact is Manchester is surrounded by lots of wildlife from coyotes, fisher, and fox to now bald eagles and peregrine falcons nesting within the city limits. Wildlife is everywhere.

5:32. A pair of geese is headed my way following the brook from the north. Immediately, another pair announces its presence to this couple and I have stereo geese going in what, a few minutes ago, was dead silence. This eruption has seemingly set the woods' alarm clock off for now crows are calling as well. A catbird and doves soon add their tones to the unrest. Canada geese have swarmed over New Hampshire during the last decade. An urban geese population spread north from Massachusetts in the late eighties bringing geese in numbers to Nashua. Geese have quickly successfully spread northward. The state now has over thirty thousand resident, mostly urban, geese. Manchester and the surrounding towns host hundreds of the geese during much of the year.

It is amazing how dynamic wildlife can be. Coyotes spread north to south from the Canadian border to all of the state in the 1970s. Opossums moved up from the south like the geese as have turkey vultures, mockingbirds, house finches, tufted titmice, cardinals, and other birds. Perhaps it was the urbanization and the food provided at winter bird feeders that have contributed to such changes. Somehow while houses, factories, malls, and even highways were being built, wildlife was filtering in around us, frequently taking advantage of humans. Raccoons, skunks, opossums, crows, and coyotes have all enjoyed the spoils of our excesses like our overloaded trashcans. All of these animals are now common well within the city limits of Manchester. Most of the increase in these animal numbers has occurred in the last two decades, just as the human population was increasing as well.

5:34. A wood thrush pipes his melody to me.

5:48. Five crows try sneaking by; only the wind rushing through their wings betrays their passage.

6:10. Another outbreak of dueling geese.

6:14. I flush a mallard drake near a recently capped landfill. One of the pairs of geese stands boldly atop the grassy knoll of the landfill cap and silently watches as I make my way by trying not to disturb them.

6:17. A chickadee startles me with a scolding, inches from my head. I like chickadees.

6:36. I have made my way southward from the old trolley line along the capped landfill and I am now approaching an old beaver dam where usually something lies waiting. A series of movements in the water has captured my attention. Aha! It's a couple of painted turtles frantically swimming about. I check my GPS unit and record the position: N 42 degrees, 53 minutes, 36.1 seconds and W 71 degrees, 32 minutes, 50.3 seconds. I have been recording sightings of reptiles and amphibians for a decade or more as part of a statewide effort to learn about their distribution and abundance. While these painted turtles are relatively common, and certainly not rare, long-term scientific observations are needed to gauge just how common they are and, hopefully, any changes will be detected in their abundance. These are things I'm always looking for anyway, so why not incorporate them into a database? In fact, I encountered a rare spotted turtle in the brook below the beaver dam a few years ago and I found a shell from a rarer wood turtle, not far from here, a while ago too. As I stare a while longer, I notice more and more telltale signs of turtles in the pond. The turtles are restless this morning. Why, I wonder?

The capped landfill made me realize how, over the last several decades, we humans have taken steps to clean up the environment and promote the wellbeing of the wildlife and fish in the Manchester area.

Absolutely none is more important in my mind than the turnaround that occurred in the Merrimack River beginning in the seventies. When I grew up in Londonderry the Merrimack River was an open flowing sewer, literally. You could not stand on the sloping banking or any rocks in the river because they were covered with a slimy coating of sewage. I was fortunate to have been part of the cleanup. I worked on various construction crews from 1969 until I graduated from college in 1972. Part of my work was helping to install the piping for the Manchester sewage treatment plant. By the mid-1970s the Merrimack River had taken on a whole new face, a bright one at that!

The addition of a fish ladder at the Amoskeag Dam by the late 1980s was the crowning glory to a river that deserves no less. Each year I make it a point to stop in at the fish ladder to witness the marvel of river herring ascending the ladder. The best part of the experience each year is to watch the faces and reactions of the other kids (and believe me, if you visit you will feel like a kid) as they experience the magic of life in the river. A dozen bald eagles regularly winter along the Merrimack and this spring the biggest news was a pair constructing a nest along the riverbank. A pair of peregrine falcons has built a nest this year within sight of the Amoskeag Dam as well.

7:15. I have nearly completed my loop of the wetlands on this plot. A yellow warbler calls from the brush along the brook. This is the first one I've heard this year. The sun now brilliantly backlights the gorgeous new green leaves in the canopy above my head. This is a glorious day, a great day to be alive in May.

7:35. The day has warmed quickly. The thermometer that hangs from my jacket zipper reads over 70 degrees. While the warmth from the jacket is no longer appreciated, it is staving off the swarm of black flies that has found me and has enveloped my head. It's funny how some things just never change. Though I have lingered here longer today than usual I now make my way full circle to the trolley bed and can see my truck in the distance. A welcomed sight at this point.

*The fish ladder at the Amoskeag Dam is still functioning. There are plans to build a new fish ladder at the Hooksett Falls Dam in 2024. This has been delayed from 2023, due to high waters.*

*First published in May 2001.*

# My Tree Stand, My Castle

Why do perfectly normal, and some not so normal, grown men hunt from tree stands? The answer is obvious, albeit somewhat complex.

My urge to hunt from a tree started very early when I was just hunting for things to see. I spent a good part of my preteen years growing up on a dairy farm in northern Maine. The farm that my dad owned was blessed with stacks of boards we had salvaged from an old shed we had torn down soon after our arrival on the farm. Boards, trees, and kids placed in close proximity on a farm can only lead to one thing—a treehouse (stand). Fairly soon treehouses began to grace the stout box elders that flourished in the fertile ground around the remaining barn that my father refurbished to contain his prized herd of registered Guernseys.

A treehouse was a perfect place for a boy to while away an afternoon before chores. In fact, my older brother Alan and I constructed several treehouses. Some together and some as our own, which we fiercely guarded. (I remember keeping a box of crackers and a jar of peanut butter in one of mine.) Even a double-decker sprouted from the crotch of the hapless tree to the east of the barn. This one peered out over the vastness of the apple orchard with the gently sloping pasture beyond sweeping down to the "swamp" that was another of my favorite places. I bet the old trees looked like a great blue heron rookery with nests or houses awkwardly jutting out from each major crotch in the trees. Some were square, one was rectangle (the most spacious), and even a triangle or two squatted amongst the tree limbs.

From a treehouse (stand), the world spreads out in a radius from your position like spokes from the hub of a wheel. You are the master of the particular piece of good earth that you command from the view. Like a treehouse, a tree stand really is your castle from which you command a view of a vast empire whether real or imagined. One only needs to look at the castles of Europe or the Buddhist temples in Asia to sense that you are part of a longstanding tradition. These castles are all built on the pinnacles of the hills and mountains for the occupants to oversee their domain.

Oh, what a life you lead high in a tree! A deer stand is no different than the castles and treehouses of yore. The vantage point sparks the imagination! Each squirrel, bird, snap of a twig or rustle of leaves stirred by a distant breeze stirs the imagination as well. Is it a big buck headed your way? A hunter strains his senses of listening and watching, his eyes focused to catch the slightest movement trying to detect the perpetrator of the sound. A fluttering leaf is carefully examined to make sure it is not the flicking ear of a deer. The horizontal limb on the distant tree suddenly looks like the top of a deer's back even though you have stared at that same limb a hundred times before. A hunter's heart races and leaps into his now dry throat. But alas it was only the wind, so the hunter tries to relax until the next subtle noise startles him once again.

Modern tree stands, like our treehouses as kids, come in all shapes, sizes, and manner of attachment to a tree. All are designed to connect you to a tree amongst its branches and by doing so it connects you to the wild world about you. A hunter can become one with nature if he applies a measure of patience.

Anyone who has spent any amount of time in a tree stand can attest to the feeling of closeness to nature. It takes at least twenty minutes to an hour before the subtleties of the wild world around you begin to reveal themselves. But they do. First you begin to hear all the noises of the wild, a scampering chipmunk or squirrel, a blue jay scolding. Something just beyond your vision. How can so many things be going on right before your very eyes and ears when only a half hour ago when you walked to the tree stand the woods seemed to be devoid of life as your feet trod on the seemingly lifeless ground? The difference is from a tree stand you are no longer interfering with the flow of life that is natural to these woods. By being quiet, motionless, and lifting your scent off the forest floor you have returned the woods to their natural state. A person walking through the woods sends out a wake of disturbance like a boat or a canoe on a tranquil pond. Your ripples are sensed by the wildlife and they flee or hide. A tree stand removes your ripple from the forest floor.

The longer you remain quiet and still in a tree stand, the more life in the woods reveals itself. You can begin to sense the very pulse of the earth and the flow of life through it. A tree stand is your castle. A place to gaze over the land and places you love, and the moments you learn to cherish as they become lodged in your memory forever.

*First published in 2003.*

# The Slippery Slimy Scary American Eel!

Eels are one of the most intriguing animals I know. They are one of the best creatures to scare a girl with. I learned that over four decades ago. For a youngster, going eel fishing was as good as you could get. No matter how you looked at it, eel fishing was always an adventure. Fortunately, some things, irrespective of time, just don't change!

So whether you are thirteen going on fourteen or forty-eight going on fourteen, eel fishing can bring out the best of the boy in you. I have had a good number of "eelin" adventures over my nearly five decades of existence. Let me share some of them with you.

My first recollection of eel fishing was when I was six years old and lived in northern Maine in "The County," Aroostook County, that is. At dusk an old trapper would settle in along the banks of the Prestile Stream in Easton not a quarter a mile from where I lived. I soon learned that he welcomed my stares and a couple of nights later he let me bring my fishing pole down for some "eelin." He even baited my hook with his special bait so I could help him fill his bucket with eels. He claimed that eels made the best mink bait going. So the late summer evenings were spent gathering bait for his winter trap line.

Later, in my early teen years, I would fish off the end of a boat dock at camp in our Maine hometown. Eel fishing can be a moment devoid of a sense of time, a Twilight Zone. I liked to catch a few perch in the afternoon and have them all cut up and ready for the arrival of darkness. A late summer's night was still the perfect time to fish. The cool nights nearly always brought a coating of fog to the water that lapped at my legs as they dangled from the dock. The bullfrog's calls resonated in the stillness of the tranquil darkness interrupting the gentle hiss of the Coleman lantern. The sudden splash of a frog or a fish that shattered the stillness sent ripples at me through the fog from the unseen event. When you are eel fishing this invariably sends another ripple, a ripple of goose bumps that emanate from the standing hair on the back of your neck, which then resonate as a prickly feeling on the backs of your legs. Eel fishing connects you to the earth like no other fishing.

One day while bass fishing with my uncle Dick Orff on this same lake, I found a hand line lying in the bottom of the old wooden boat as we lay at anchor off Turtle Island. A small yellow perch was quickly added to the end of the line and tossed overboard. A few minutes later a steady tug on the line meant something was going to happen. Sure enough, during the middle of the day I yanked in the biggest eel I ever caught. About that time my uncle was standing (make that dancing) on his seat as the eel slithered beneath it! The eel measured three feet long and weighed four and a half pounds! A trophy for sure.

Sometimes you can just get plain lucky with eels. In 1971, when I was a student at UNH, my lifelong fishing partner was home on leave from the Navy. Rick Hamlett had just returned from a tour of duty on a destroyer along the coast of Vietnam. We made up for his long absence from fishing by not wasting a minute of time, day or night. This particular night found us checking out a pond near Durham that had a reputation for its bass. To our amazement we found the face of the dam creating the pond covered with juvenile eels shimmying up the near vertical surface to the pond above. We knew we had struck gold. No bass could possibly resist these three- to four-inch-long delicacies. So using a shirt for a net we scooped up a bucketful. Back to my apartment in Exeter we went to get fishing gear.

We soon realized that we had too many eels to keep alive so I carefully transferred the extras into my ten-gallon aquarium. They turned it into a black writhing mass! We carefully and quietly gathered the necessary fishing gear so as not to awaken my wife and headed back to the pond. When my young bride woke up and got ready to walk to work nearby, she was somewhat startled by the caldron of eels I had left in our living room.

She was even more startled by their absence in the aquarium when she returned for lunch, four hours later. What luck, they had all crawled out! Our lovely apartment had been transformed into a scene from an Alfred Hitchcock movie. They slithered everywhere! Did I mention girls are scared of eels? They are! Luckily a neighbor came to the rescue and collected many of them. Some had made it through our kitchen all the way to the bedroom. Eels are amazing survivors. So am I!

So if you are looking for a little adventure this summer maybe you should consider a nice tranquil evening of eel fishing. It won't stay tranquil for long once you hook an eel. Be sure to carefully release your eels, or better yet, cut the line as close to their mouth as possible. Releasing an eel is an investment in your future, a time you are thirteen again!

*First published in 1993.*

# Me and Three Bears Up a Tree

I got the call in the late evening at home, as was common: a bear problem. (This actually occurred on June 26, 1981, as I pulled directly from my diary.) However, this one was different than most. At the Balsams Resort in Dixville Notch, a troublesome bear had just gone into the culvert trap set by the local conservation officer and was caught. Trouble is, there now were three bear cubs wandering around the trap! Female bears with cubs were seldom problem bears, but I had an exception on my hands.

I was troubled as to what to do. Call the conservation officer and just let the mother bear out? Take her away and let the cubs fend for themselves? Or maybe, just maybe, catch the cubs too and move the whole family.

If I could catch the whole family unit, I already had a plan to release them into the southwestern part of the state where few bears lived. The southwest corner of the state holds exceptionally good black bear habitat, but bears had been killed off from this area by the mid-1800s. Although bears were gradually expanding their range back to this region, a mother bear and cubs could restore a population sooner.

I decided to go for capturing the whole family, so I called Bruce Cairns of Jefferson, the president of the New Hampshire Bear Hunters Association, to see if I could enlist a pack of bear hounds to locate the cubs the next day. As expected, I got the nod from Bruce. He would take a day off from his job the next day to help capture the cubs with his hounds, if needed.

It was mid-morning the next day before we gathered at the scene with hounds, ropes, a big net and tranquilizer equipment. We tried a search of the surrounding steep, beech-tree-covered slopes to no avail. The three cubs had disappeared! It was time for the trained noses of the bear hounds to locate the cubs.

And they did, not twenty minutes later, a few hundred yards up the steep slope behind the resort. The cubs were just dots of black hidden in the shadows of a seventy-foot-tall beech tree. No wonder we could not find them. We were hopeless without the trained hounds at our disposal. We all stood around scratching our heads trying to figure a way to get the cubs safely down out of their lofty perch.

I volunteered to climb the huge beech tree with the tranquilizer gun and tranquilize each cub, which then was to be caught in the ten-foot by ten-foot cargo net held by the others below. To make matters worse it was a very steep slope that was covered with boulders, making it difficult to even stand and hold the net taut.

I managed to shimmy up the tree trunk and slowly gained my distance on the closest cub so I could safely hit it in the rump with a small dart. Of course the cubs all scrambled even further up and out to ever-thinner limbs, making the shots more difficult.

*Pow-thump*, the first dart hit home, causing the little cub to scamper even more. I frantically commanded those below to maneuver this way and that, trying to line the net far below with the trajectory of where I predicted the cub would fall. But it kept moving and I kept yelling. Then the cub gradually started to fall, only to catch itself and move again. Finally, after an endless wait, it tumbled to those waiting below. Into the net it went, but only into one corner, as it ricocheted off the net and cartwheeled down the steep slope. Biologist Henry Laramie vaulted down the slope after it. When Henry checked her, she seemed to be okay. Yes, a female cub, just what was needed to bolster a fledgling bear population down south. Henry was worried about her so he lugged her back to our pickup and placed her on the seat in the cab, as we expected the tranquilized cub to be out for a while before she recovered.

Twice more the scene was repeated with me climbing even further up the huge tree and out onto too-thin branches to make the shots as close and as safely as possible. By now the ground crew had perfected the capturing in the net, safely scooping the next two cubs out of the air. Further great news, they too were both female cubs. We now had four female bears to release!

Down off the hill we clambered after the successful capture of the last two cubs. We were anxious to reintroduce the cubs to their mother, who had entered the trap over a half day before. But this is where things got rather interesting again. Trying to open the bear culvert trap up enough to put a cub in, and not let the mother out, was not very easy. The cubs were wide awake and were now fighting and clawing us. We had struggling, fighting cubs on one side of the door and a rather upset mother bear on the other side of the door. The mother bear luckily snatched the fighting cubs out of our fingers as we tried to stuff them in. Each time we carefully examined our hands to make sure our fingers *were* still attached!

Once we got the last two cubs captured into the culvert with Mom, we turned our attention to the cub in the cab. Not a pretty sight. This cub had wakened some time before and was tearing the cab to shreds. It sure looked like she would be long gone again if we opened the door. But we managed to nab her and settle her in with the others after a feisty tussle.

We were all wearing rather happy grins as we drove back south late that afternoon. I kept all four bears in a shady area at my home that night with plans to head out early the next morning to the release spot.

I chose Pitcher Mountain in Stoddard for the release site. This region had excellent bear habitat and was twenty miles south of where I felt the resident bear range ended. Our good fortune continued as we contacted a professor at the University of Massachusetts who had a graduate student just beginning a bear study there. They wanted to "test" one of their new radio collars on a bear, so I offered to place one on the mother bear before we released them all.

Early the next morning I met graduate student Ken Elowe. I tranquilized the mother bear and we tagged all the bears and fitted Mom with a radio collar. Once she was fully awake, we opened the culvert door and watched all four bears scamper off into their new home. Ken's radio tracking over the next weeks and months showed us that she pretty well stayed near where she was released. Females without cubs and male bears have a great tendency to return back to their capture site. We have had other bears return over a hundred miles in New Hampshire. But the young cubs acted as an anchor, keeping this whole family in what was now resident bear range. The reintroduction worked!

*Note*: The black bear population was estimated to be about twelve hundred to fifteen hundred bears in the early to mid-1980s. The legislature granted the authority to the Fish and Game Department to regulate bear hunting in 1985. In 1986, the southern half of the state was closed to bear hunting. In some small part, thanks to these females, the bear population grew and twenty years later is estimated to be about five thousand.

*As of 2024, there are now over six thousand bear spread over the entire state. The Monadnock region continues to be well populated.*

*First published in 1990.*

# May I Have This Dance?
# A Waltz With Fire

Fish and Game biologist Ed Robinson's face and neck appeared red and sunburned as I glanced at him mid-morning. It was not the fire of the sky but the fire of scorched earth that had transformed Ed's boyish face into a reddened sweating orb as he continued to dance with the fire consuming the brush, grass, blueberry bushes, and solitary white pines in Bear Brook State Park one April day. Ed is a member of the state's "Fire Crew."

This fire crew actually starts fires each spring in order to create and maintain habitat suitable for a host of species, especially those that need to feed, nest, or use open areas in the forests. The majority of the fire crew are from the state's Division of Forest and Lands and some are forest rangers whose job is normally preventing or putting out fires. Frequently, local town firefighters join the crew when the prescribed burn is in their town.

Prescribed burns are fires used to achieve specific goals for habitat management at each site. The crew's first fire was at the State Forest Nursery in Boscawen in 1990 and was designed to burn the understory (brush) in order to promote the red oak seedlings. Since then other prescribed burns have been done at the Hopkinton-Everett Flood Control Area to maintain fields, Bear Brook State Park to restore blueberry fields, Pine River State Forest to regenerate and maintain a unique pitch pine forest, and at Blue Job State Forest to maintain blueberry fields and open areas. Prior to each burn, a "burn plan" is developed that clearly defines why fire is the best and most economical tool to provide the desired results.

The prescribed burns help rejuvenate the forest openings preferred by animals that eat the grass and blueberries or nest in the open areas. More and more of the once-open landscape in New Hampshire has been choked by the advancing forests. Our state is now over 83 percent forested. Our state's forests have grown old and openings in the forest canopy have

*The state fire crew briefing on the fire plan for that day.*

vanished. Fire rejuvenates the forest. Some areas like Bear Brook State Park have been burned twice. After each fire in the park the high use of deer browsing the blueberry bushes the previous winter is evident. Numerous smoldering piles of deer droppings persisted long after the other fires had died out. Several songbird species prefer the openings created in the high forest canopy as well as snowshoe hare and woodcock.

There was a good reason why Ed's face was reddened, as controlling a prescribed burn is like a dance with the fire itself. Most crew members wear a backpack water tank equipped with a sprayer. In Bear Brook the fire is literally weaved around the relatively small openings in the white pine forest. The sandy soil and dry conditions needed to successfully burn also means that the fire is anxious to spread into the forest. Fire Ranger Brian Nowell, whose stout stature clearly would lead you to believe he is a descendant of Smokey Bear, carefully waves the drip torch, which spills out a flaming mixture onto the ground. Each opening is carefully lit, beginning on the outside of the opening, so that the fire tends to race inward and extinguishes itself when it draws together at the center just when it's the most intense. The fire crew stands guard on the edge of the trees and carefully puts out the fire spreading outward from the circle before it reaches the trees. This is a face in the fire job. Crew members are bent over reaching into the base of the flames as they spread outward, spraying here to squelch the flames while letting them advance there. Sometimes your pant legs become nearly unbearably hot! But you can't step back; you must hold your ground and maintain control of the fire's leading edge.

Each crew member must dance with the fire. Usually, it is a steady waltz. But sometimes an unexpected breeze, often created by the fire itself, picks up the tempo! A fire devil may whirl down the fire line, like a pair of polka dancers at their granddaughter's wedding. Surprising, yet inspiring, as it whirls and you dance along at least with your eyes and mind. You never let fire step on your toes; you must move in to take the lead with this partner! A dance with fire always leaves your face and any exposed skin reddened and your palms sweaty. The fire crew is anxious to continue to help the wildlife of the state. May I have this dance?

*Prescribed burns to help wildlife continue to this day.*
*First published in 1999.*

# Diving Into the World of the Dead

Conservation Officer Jim Kneeland patiently stood at the icy precipice, staring into the inky black hole, the entrance to the world of the dead. He was duty bound to enter the world of the dead, he had been trained for this very moment, yet in the last few moments before donning his remaining diving gear and plunging into the abyss, he gleaned the senses of the world of the living around him. The wind was bitter cold as it whipped particles of snow and ice down the expanse of the frozen lake with a distant roar. The icy crystals bit at his exposed cheek, so he turned his face toward the sun, the light of life, to warm the flesh despite the coldness of the twenty-degree air. How different were the sounds of the murmuring voices gathered on the ice in an arc around the hole he was about to enter. His mind flashed back to moments he had spent patrolling this lake on a warm summer's eve. The sounds of laughter shrieking from a young water skier above the din of an outboard motor echoed in his memory. Summer wind could be as warm and gentle as a grandmother's kiss on your cheek; so unlike the fierceness of this winter wind. He donned his mask and tested his air supply shutting out the sounds of life, but the brilliant sun still reflected in the glass plate that would soon protect his face from the near frozen water he was about to enter.

Jim had volunteered for the New Hampshire Fish and Game Department's dive team six years before. He was now an experienced diver but the challenges of ice diving tested his skills. Unfortunately, ice dives most frequently are the result of a drowning hours before. There is no rescue; the best that can be hoped for is the quick recovery of an unfortunate victim. The Fish and Game Department is usually responsible for the recovery of fifteen to twenty-five victims a year. About a third of them are winter recoveries. The close-knit team of divers are called upon all too frequently for a job they solemnly must do, a job that, to a man, they wished they didn't have to do. Jim remembers that when he first joined the team he spent near sleepless nights before a recovery dive, pondering the next day's task. Experience has served him well and his mental anguish has been tempered by the knowledge of the appreciation of the families who have had their lost ones returned. Jim intentionally stays aloof of the family or friends gathered nearby. The dive must remain emotionless, professional to the highest degree, focused on the job at hand in order to reduce the risk of loss of his life or his partners.

*NH Fish and Game Conservation Officer James Kneeland.*

Jim gently slipped into the gaping hole in the ice, quickly followed by his diver partner Conservation Officer John Whitemore, another experienced diver. Jim's anxiety level grew as the lead-filled weight belt around his waist accelerated his drop into the world of the dead. The inky blackness of the frigid water swallowed him as surely as it swallowed the victim the day before. The sense of death was at hand. The frigid water would gladly and swiftly sap the very warmth of his life if given just the slightest chance. The frigid water pressed ever harder against his diving dry suit as the water pressure increased with depth. He was amazed at how quickly the halo of light from the surface was extinguished by inky water as he descended into the world of the dead.

Suddenly, his descent was slowed but not stopped by the bottom. His feet slowly glided into a deep muck lying in wait at the bottom of the lake and his descent did not stop until the muck was nearly to his knees.

Darkness, absolute darkness enveloped Jim. Jim reached for his gauges to read the depth, only to find his gauges unreadable even when he pressed them against his faceplate. Only the faint glow of the luminous dials was visible. The darkness of death had isolated him from the rest of the living world. He extended his arms out into the blackness, groping for what lay in the total darkness. Frequently he has found the victim directly below the hole in the ice, lying on the bottom. This time he was pleasantly rewarded by the feel of his dive partner, less than an arm's length away but totally lost in the blackness.

A search of the bottom near where they had landed turned up nothing! Not even the vehicle the victim was riding on the ice directly above. Jim had been warned by other divers about the lack of visibility in this lake; one had nicknamed it "the nightmare lake." This was darkness that he had never experienced before. Complete! Jim frequently conferred, via his special underwater radio, with Dive Master Tim McClare directing the dive from his position above. Jim felt the reassuring tugs on his lifeline to the crew above. In every sense of the word it was his lifeline, as sure as an umbilical cord. He commanded to the crew above to keep his lifeline taut, lest it droop and become ensnared in the vehicle or other debris laying somewhere close at hand in the darkness. The sense of death was at hand; a kick of his partner's flipper could rip his face mask off in a second and sever his lifegiving air supply. Death lurked in the absolute darkness; he, searching for death and death searching for him.

His heart grew heavy as he realized that this time it might take days of diving in this blackness to recover the body. A body that could even have been swallowed by the thick layer of cold muck.

After a futile search and with their safety in mind, he and his partner began to ascend toward the surface. The tug of the lifeline was ever so reassuring as the faint glow of the underside of the ice became visible, a welcomed sight. Jim caught sight of something suspended fifteen feet below the ice, a pair of goggles. Could the victim be suspended too, he wondered?

Fitted with a new tank of air, Jim reentered the dark water to do a search just under the ice, sweeping in a circle around the hole. First about ten feet of rope glided through the hands of John, who stayed at the hole tethered to Jim so that a search pattern could be done systematically. Chunks of fractured ice, from the event of the day before, hovered under the ice, distorting Jim's view and threatened to snag his safety line. His commands brought more rope to him as he slowly expanded his search, twenty, then thirty and forty feet.

Just at the edge of his limits of sight, he recognized the body of the victim floating up against the ice. A strange sense of relief swept over Jim as he grasped the victim's clothes and gave the signal to pull them in. His lifeline grew tight as John pulled, reassuring Jim of his return to the world of the living. The sunlight felt especially good to Jim as he removed his mask in the tranquility of a frozen moment.

*First published in 2002.*

# What is a Hunter?

A hunter is difficult to define. Sociologists have written volumes in attempting to define and portray a hunter. I'm no sociologist, but I am a hunter. One thing most hunters possess, and I would say all successful hunters, is the ability to observe the ecosystem around them. Both plants and animals.

There is no single definition of a hunter because, in one fashion or another, we are all hunters and we are all unique. You don't need a bow or gun to be a hunter; a shopping cart will do. But, I'll focus my definition on those who hunt wild game to eat rather than bagging it at the meat counter.

Instead of trying to define what a hunter is, I think a great way for you to visualize a hunter is to describe two that I have known during the thirty-five years that I have considered myself a hunter.

## Carl

I first met Carl in 1964 when I was fourteen. I took the NRA hunter safety course sponsored and taught by members of the Londonderry Fish and Game Club. Carl was a member and soon took me under his wing. He taught me much about wildlife and, more importantly, hunting ethics over the next decade and a half until his death. He also became my bridge to the past.

Carl was a fox hunter though it had been decades since he had killed a fox. The sound of his Walker hounds chasing a fox through the wilds of Londonderry rejuvenated his spirit, which was obvious in his physical transformation as he sat on the tailgate of his Scout. Each Sunday he would be parked along Pettengill Road, he called it the old name "Skim Milk Road." This road is perpendicular to the south end of the Manchester Airport north-south runway. The Pettengill cemetery monument is plainly visible from the new terminal, but back in the sixties and seventies a lush canopy of hardwoods and mixed pine enveloped this fourteen-hundred-acre tract of land now being devoured by industrial development. It was a magical place back then.

Though he was "on in years" in the 1960s and physically unable to keep up with his fox hounds as they coursed the vast woodlands with an occasional yelp, his mind's eye followed each leap and bound of the hounds. He knew these woods well! He reminisced about hunting and trapping in the twenties and thirties when fox hunting was at its peak in the state. There were dozens of hound packs back then. He passed along the heritage of knowing and understanding the land and its wildlife. He always extolled the need to "put something back" . . . and he always did. He was an active club member and was quick to raise his hand for a motion to buy wildlife shrubs to plant, pheasants to stock, or to send a delegate to the state house to testify on behalf of a bill to benefit wildlife. Carl was a doer.

Carl had his own pet projects, too. I remember finding small galvanized containers around the hedge rows in Londonderry, Litchfield, and Hudson. This was one of Carl's winter projects. He would buy crack corn with his own money each winter to "help" out the pheasants and other critters. Carl was always putting something back. So, I have always tried to follow in Carl's footsteps as a wildlife biologist for the New Hampshire Fish and Game Department. I always try to put something back: to restore or improve a wildlife species, to bring the ecosystem back into balance like Carl. Carl's spirit was with me a few years ago as I opened the crates releasing over twenty turkeys into Chester. A town that had not heard the gobble of an old tom in one hundred and fifty years. Carl's spirit soared on turkey wings that day. He has been with me as I cast thousands of Atlantic salmon fry into the Lamprey, Isinglass, and Suncook Rivers. Or even the day when a helpless cow moose was trapped in a backyard on Elm Street in Manchester; I tranquilized her and moved her out of the city. Carl would have loved to have seen her gallop away. I think perhaps he did. She loved being "put back."

Hunters like Carl have always supported, both financially and willingly, these restoration efforts. Thank you for these valuable lessons, Carl Suosso.

## Bill

Here is a hunter with a great power of observation with skills of patience and determination like no other person I have ever met. Bill gets things done and he usually gets his deer. Bill was one of my hunter safety instructors and a lifelong friend. Bill is a man who can focus his skills, observation, patience, and determination into any project. When deer season rolls around, his focus is deer. Some people would categorize Bill as a disabled Korean War veteran. The number plate on his vehicle would corroborate that. But, from my experience, you would be hard put to use the word *disabled* in any description of Bill. Try winning an argument with him. He is astute in his knowledge of a vast array of subjects. He does his homework. He is a keen observer of life.

He was the club's legislative liaison for several years and later served the town of Londonderry as a representative for over two decades. As a hunter and more important, as a conservationist, he has sponsored or shepherded numerous bills through the legislative process. Plan to use your muzzle loading rifle in a "shotgun only" town this fall? It was Bill's "bill" that makes it possible. How about launching that new boat of yours at one of the state's new access sites? Bill led the charge on that one, too. After a couple years or more of me hounding Bill to take action, Bill introduced legislation that was passed, creating the Fish and Game Access Program. Speaking of access, have you ever noticed more and more buildings have been made handicap accessible in the last two decades? Bill is a hunter of more than deer; he is a hunter of equality as well. Ask any legislator about Bill Boucher: they will remark, "Oh, the guy in the wheelchair." Everyone knows Bill doesn't just step on your toes, he'll roll right over you if you are in the way of progress, or his deer.

Hunters are men and women with a sense of connection to the environment. Hunters have financed the return and re-establishment of numerous wildlife species including moose, deer, bear, and turkeys. Hunters turned the tide against misuses of our natural resources and have brought about the golden years of wildlife in New Hampshire. Hunters possess a deep and abiding respect for nature and seek to pass this heritage on to the next generation by setting positive examples throughout the years. A hunter is a conservationist savoring the abundance of wildlife today and ensuring its survival into the next millennium.

*First published in July 1998.*

# Henry . . . Just Henry

On June 6, 2002, I lost a friend to cancer: a former supervisor, a mentor, and someone I knew to be a really true conservationist, wildlife biologist Henry Laramie of Pembroke (1926–2002). I knew Henry for about thirty of his seventy-six years. Henry was my supervisor until his retirement from the New Hampshire Fish and Game Department in 1985. His career spanned forty years there. Conservation enveloped Henry's life to the very end. He was frequently noted as "Henry from the Fish and Game Department." Or just "Henry" by most biologists or conservationists. There are few people who earn the privilege of just one name. A first name at that! Henry was one. Let me tell you about "Henry," the friend I know:

The first time I remember meeting Henry was in the early 1970s, while I was duck hunting in Little Cohas Marsh, a Fish and Game Wildlife management area in Londonderry where I grew up. I had known this area for a decade since my family moved to Londonderry in 1962 when I was twelve. Henry helped create this marsh, or "swamp" as my mother and most others would call these wetlands. Henry pioneered the program that led to the construction of a dam that created Little Cohas Marsh and many others around the state beginning in the 1950s. Henry, ever a contrarian, recognized these "swamps" for their wildlife value, contrary to most people in that era. He was instrumental in creating marshes such as Hayes Marsh and Hall Mountain Marsh in Allenstown. Doles, Casalis, Burnham, and Woodman marshes are also the results of his efforts. They are bursting with ducks, herons, turtles, and frogs to this very day, thanks to his foresight.

Henry nearly singlehandedly led the most productive land protection efforts in the state for two decades in the seventies and eighties when there was no funding at the department or elsewhere for land conservation.

*Eric Orff left, Henry Laramie, right, tagging bear cubs.*

Henry was a finagler when it came to buying and protecting wildlife habitat. Using just federal dollars and no state funding Henry acquired over twelve thousand acres of habitat, which are now protected in perpetuity, including the department's largest holding of over four thousand acres, Enfield Wildlife Management Area (WMA). Over half of the land protected by the department up to the time of his retirement was Henry's doing. And that really was not even part of his job. He had lots of other duties as a supervisor. It was just something else that Henry managed to accomplish, without any actual intended funding, in his spare time. Henry worked miracles at Fish and Game. His absolute honesty and relentless commitment to conservation has left a legacy in this state of protected wildlife habitat to be enjoyed by generations yet to be born. They include: the Enfield WMA, over four thousand acres; Leonard WMA and Knights Meadow Marsh, over a thousand acres; McDaniels Marsh, three hundred acres; Kearsarge WMA, a thousand acres; and the Brown Company lands in Pittsburg, over a thousand acres. There even is a Laramie's Marsh, rightfully so.

Bears were another favorite of Henry. He began trapping, tranquilizing, and moving nuisance bear in New Hampshire in the early 1960s. Bears could be killed year round and were not protected by a fall-only hunting season until the mid-1960s. Again Henry showed his foresight in trying to change the image and value of bear by wanting to conserve them. The state had paid a bounty on them until 1956. I bet Henry had a lot to do with removing the bounty.

Well, Henry's perseverance paid off over time. In 1978 Henry was able to finally get funding for a bear project at Fish and Game. I was fortunate enough to be hired by Henry to head up the bear project. This is where I really began to learn the ropes from Henry. Henry was a masterful teacher and a great mentor. Henry taught by showing you how to do it and explaining that he knew this was a good way to do something because he had tried all the other ways and had failed. Henry was not afraid of failure; he was just afraid of people failing to try their hardest. Henry taught me that failure means you are trying your hardest and being innovative. Just never give up because Henry never did. New Hampshire now has over five thousand bears thanks to Henry's leadership and foresight.

Henry was always a thinker and tinkerer. What he didn't get done at work he simply took home and worked over the weekend on it. I remember he worked weeks trying to perfect a jab stick to tranquilize animals with. Bear culvert traps, beaver pipes, radio collars, bear tranquilizers, deer trapping, and duck boxes, to name a few, are things that Henry invented or improved upon.

Henry knew this state's wildlife intimately and shared his lifelong passion for wildlife with whoever he could. Even after his retirement Henry stayed active as a conservationist. He hardly ever missed a Fish and Game Commission meeting and the last couple of years he became more active in the New Hampshire Wildlife Federation. Recently he served as a director as well. When you catch a glimpse of a bear, a turkey, or a moose, think of Henry. Henry sparked the notion that led the way for their return. Henry left a lasting legacy of wildlife conservation in New Hampshire. Henry made a difference in the number of bears, moose, turkeys, beavers, and scores of other wild critters in New Hampshire. Henry made a difference in thousands of acres of protected lands for conservation in New Hampshire. Henry made a difference in thousands of acres of wetlands created and protected in New Hampshire.

Henry . . . made a difference in me.

*First published in June 2002.*

# With a Cluck-Cluck Here, and a Gobble-Gobble There

*Old Ted Walski Scattered Some Turkeys In New Hampshire: E-I, E-I, OH, OH, OH!*

New Hampshire's turkey population has grown to an estimated twenty-five thousand birds from an initial transplant of only twenty-five wild turkeys transferred from New York in 1975 and released along the Connecticut River in Walpole. New Hampshire Fish and Game turkey biologist Ted Walski nurtured the fledgling flock like a mother hen from nearly the very beginning. Ted spent countless hours monitoring the turkeys, then capturing and transplanting some to new sites across the state to extend their range. All the while he educated anyone who would listen about his favorite subject, his turkeys.

Indeed, Ted has been very successful over the nearly three decades in spreading turkeys to about every corner of the state. But even he has been pleasantly surprised by what he has witnessed over the last decade. Turkeys have a mind of their own and have done what nearly every turkey hunter knows that they are capable of: doing the unexpected!

The fact is, even Ted is surprised at how adaptable some turkeys have been. Turkeys have learned to cope with us humans by living very safely in backyards in Salem, Atkinson, Hudson, and Hollis, and some have even become a nuisance in a condo complex in Pelham. There are turkeys that enjoy the sunrise over the Atlantic Ocean from their perches in North Hampton or Rye. Now there are even turkeys that have taken up residence in Pittsburg, a town along the Canadian border!

Turkeys regularly swap residences in adjoining states too. Maine, Massachusetts, and Vermont continually swap birds with us as turkeys fly across the rivers or walk across them when they are frozen. This has got to be the source of the "surprise birds" that showed up in some of the border towns such as Wakefield and Effingham to the east, and Colebrook, Columbia, and Stewartstown to the northwest. Ted says, "Even the Mount Washington Valley is home to a growing flock of turkeys." Even if you have not heard the gobble of a tom in the spring in your backyard yet, you should expect one soon.

Ted also wanted to point out the number of "city birds," as he called them. Turkeys have somehow adjusted to the bright lights of some of New Hampshire's cities. Ted noted that nearly every year in the last few years, hunters have been able to take turkeys during the spring season in Concord, Keene, and Claremont. This is not exactly where he had expected turkeys to be taken a decade ago!

New Hampshire had its first spring gobbler season in 1980. From a very limited hunt the first few years, turkey hunting has really taken off. Record spring harvests have occurred nearly every year since 1986, when the one-hundred mark was broken, to over two thousand taken in 2001. The number of turkey hunters has soared too. Nearly fourteen thousand hunters purchased a turkey hunting license in 2001.

In 2001, 2,259 turkeys were taken from 187 towns scattered across the state. This was a 20 percent increase over the previous record harvest the year before. Proposed changes in the turkey regulations for the 2003 season will see even more of the state open to turkey hunting. The liberalization of Wildlife Management Units (WMUs) open to turkey hunting in 2003 will include B, C1, and C2. The regulations for this year's turkey season, which will open on May 3, 2002, will not be affected by the proposed changes.

Ted is anticipating another extraordinary spring for New Hampshire's turkey hunters. Last spring and summer's drought conditions were perfect for the mother hens to successfully raise their broods. Ted says the brood production in 2001 was "very good" and noted that average brood count for June 2001 was 9.39 poults per hen. This was a record survival rate, as the thirteen-year average is 7.22 poults per hen. Coupled with the winter that wasn't and a plentiful supply of acorns on the ground last fall, which the turkeys have dined on all winter, there are more turkeys out there than you can shake a stick at. More than likely there will be many shaky shotgun barrels this spring as nervous hunters try to line up that perfect shot on a nerve-rattling gobbler! By the way, don't forget that you can buy your hunting license and turkey license online at the New Hampshire Fish and Game Departments' web site at www.wildlife.state.nh.us. All you need is your driver's license, last year's hunting license and a credit card ready. So even if it is 11:30 p.m. on May 2 you don't have to miss that early bird, just a few hours of sleep. Stop dreaming and get hunting for one of New Hampshire's great gobblers. There's probably one roosting within sight of your back porch right now. Maybe, just maybe, you should check *under* the porch too! Ted recommends it.

*As of 2024, there are over forty thousand turkeys in New Hampshire. First published in 2001.*

# New Hampshire Conservation Legends

Let me name a few I have known or know of in my seven-plus decades of life, mostly here in New Hampshire. Helenette Silver easily comes to mind, followed by Henry Laramie, John Lanier, Ben Kilham, John Harrigan, and Ted. Ted Walski, that is, and I can easily put his name at the top of the list.

Yes, Ted became a legend of legends.

No doubt, Ted really was, for most of his career, an ambassador, not only for the department, but for all sorts of fish and wildlife. Yes, Ted was *the* turkey biologist, but he really became the face and figure for so much more. For decades he was "The Fish and Game Department" to many.

I first remember Ted from the summer of 1969. I had entered UNH that previous fall as a wildlife student. I don't remember exactly how we met there, but on that summer day, he rolled into the driveway of my home in Londonderry with a UConn professor by the name of Mcdowel. Ted wanted to know where he could trap some cottontail rabbits near my house.

To meet Ted was to like Ted. Who could not like Ted right off? He was a friend to many. Ted was unique and amazing. Even Ted's voice was unique. Who among us here cannot easily hear Ted's voice, should we close our eyes for just a moment?

In my seven-plus decades I can recall only two or three people I have met that have what I will describe as amazing minds. Ted is right up there. In the field, point to any wild thing, plant or animal, and Ted could easily identify it, often giving you the scientific name as well.

Ted knew stuff.

Ted had learned the fabric of New Hampshire. It just seemed he knew something about everything. Yes, he had that kind of a mind that just kept learning and learning and learning. He seemed excited to keep making

*Getting a collar on an ornery goose can be a challenging task—even for seasoned biologists Ted Walski (left) and Ed Robinson.*

observations and figuring things out *his* way. I remember one year he hand-counted every single acorn that fell from a nearby oak tree. I wish I could ask Ted that number again.

*This tribute was read at Ted's memorial service in January 2024.*
*First published in 2024.*

# Catching the Spirit of the Native Brook Trout

Step by step I skulk through the alders, easing my way to the edge of the cold spring-fed brook. I stalk in the way of a native trout fisherman. I'll not throw my shadow into the brook to spook my wary prey this sunny late-spring day. Each step propels me further backwards in time: 1990 . . . 1980 . . . 1970 . . . 1961 (age eleven). The words of my father echo in my mind: "Walk softly, stay low, don't let your shadow creep onto the water, swing the worm, and gently place it just above that submerged log."

My father knew the ways of the native brook trout. Knowledge had been passed down from generations before and he had had a lifetime of practicing his skills. My father grew up in Down East Maine before the Great Depression and was the oldest of a large family typical of the times. He had perfected the art of catching the elusive native brook trout to treat the family with a delicious meal. I listened intently for years to his descriptions of the ways of the trout. Luckily, as a youngster in Northern Maine, and later as a teen in rural New Hampshire, I got to practice my skills often.

Wherever I have lived, the spirit of the native brook trout has lingered in the shadows of the quiet gentle flowing brooks, some without names. Following in my father's footsteps was natural as I sought the spirit of the trout. There is so much to learn, sense, feel, hear, and smell in the recesses of a forgotten stream. Discoveries are at every bend on a newly fished brook. A pond above a new beaver dam will surely contain the biggest trout of the day. An old mill site leaves you listening for voices from the past, lingering in the winds whispering through the pine tops. A familiar brook changes relentlessly; a new fallen tree here, a shadow there. A rain shower can completely change the mood of a brook overnight.

Native brook trout are small elusive creatures. In the spring-fed brooks I have fished all my life, the natives were usually only three to seven inches long. As a boy growing up in Maine, a six-inch ruler always dangled from

*A creel of native brook trout.*

my woven white ash fishing basket. The minimum length limit was six inches. I would catch a lot of trout before adding a dozen keepers to my basket. I lined my fishing basket with moss in anticipation of feeling the wriggling trout as I dropped them through the port in the top of the basket.

A native brook trout is the sweetest-tasting fish I know. I also think they are, by far, the most handsome. A native brook trout can only be described as the color of a beautiful sunset. Their white belly shines like the last rays of sun light on a perfect day. The reds and pinks of their fins and belly edges merge into dusky color of their sides almost unnoticed, like the last pinkish glow of the day turning into night. Wavy shadows ripple along their sides specked with brilliant red dots that catch your eye like evening stars before the transition into the blackness of night spilling down their sides from the back. And like the universe itself, the spirit of the native brook trout is unfathomable.

A trout is as quick as lightning. In a flash a trout will appear out of nowhere, grab your worm, and disappear into the roots stripping your hook of its worm and leaving you hopelessly tangled. To fish for natives, you must stay on the edge. No mistake is forgiven. Though trout prefer the deep, dark shadows of an undercut banking or log, they can hide almost anywhere. A gentle set of riffles, even in the shallowest of brooks, will do.

The spirit of the native brook trout is far more than the fish themselves. I can sense not only the spirit of the trout in the humid yet cool air beneath the alders and pines, but I sense the spirit of my father as well. Sometimes it's the distant drumming of a woodpecker, or the melodic organ-like piping of a wood thrush. Like the brook, the spirit is ever changing, always alive, but sometimes as elusive as the trout.

Last winter I summoned the courage to search my mother's garage for the old steel telescoping fishing pole my father had carefully tucked away; probably several years before his death nearly three years ago. When I found it I knew that it was time to bring it back to life. I'm sure this rod is nearly fifty years old. It was discovered about twenty years ago in the house where my father was born in 1923. He cherished the old rod as I do now. The simple black crank reel has stood the test of time. Despite its worn appearance the spool still fits snugly. The pearl white knob on the crank does show its age as it wobbles when I turn it softly: click, click, click.

Quietly I near the banking of the brook. Slowly I extend the metal rod; first the top section slides out, then the middle section, a little brown with rust. The black reel on the cork handle pays out my line in a steady click, click, click. Each click ratchets my mind backwards in time as the years click away. I am as excited as a kid as I creep forward. Slowly, slowly I swing the rod over the brook, relaxing my grip on the line and gently let the worm swing like a pendulum into the spirit of the native brook trout.

The worm gracefully landed on the edge of the shadow cast by the undercut banking. I wondered, "Was it far enough?" The telltale rattle, shimmying up the fishing line, soon set the tip of the steel pole quivering. My answer was telegraphed to me by the fish with a series of taps on the line. The taps soon transcended the rod and quivered the cork handle in my hand. I sensed the grip of my father from the last time a native trout quivered this handle. My eyes quickly glistened.

*P. S. Happy Father's Day, Dad.*

*This tribute was first published in 1999.*

# The Long Journey to the Opening Day of Trout Season

For a young trout fisherman, thinking of the opening day of trout fishing season on the fourth Saturday in April was the stuff of dreams. School days were spent daydreaming, staring out the school windows watching the shards of the icy winter slowly … ever so slowly … melt away.

The journey to opening day included at least weekly stops at the only local sports shop to spend our meager allowance, or any other money we had managed to earn or beg. Since I was in seventh grade at the Londonderry Central School, that meant a trip to Berdett's in Derry. This bob-house-sized shop was crammed with lures and fishing paraphernalia my fishing buddy Rick Hamlett and I hoped to buy to fill our tackle boxes. We intimately knew each lure as we had fondled them well before any purchase. Some evenings during the cold winter we would get together to go over our tackle boxes and trade lures. This meant trying to trade off a lure I had found the previous summer for one he had bought. So, not only did I need to keep track of every lure and gadget in my tackle box, but I had to have a reasonable knowledge of his!

Eventually the magic of opening day arrived. The pent-up anxiety gushed as we frantically made the final preparation. Worms were dug; rods, reels, and lines were checked and rechecked. The excitement flowed like the spring rapids we would soon challenge. Burbles of laughter were echoed by the high pitches of our preteen boyish voices. Back in the 1960s, all trout fishing started the fourth Saturday in April, so we knew there would be competition for the best spots on the brook.

Rick Hamlett and I always "camped out" the night before the opening morning. In our teen years that meant his backyard or mine, due to popular demand of our parents. Of course Londonderry was very rural back then, so it truly was a wilderness experience.

I remember camping out in his backyard in April 1963. We decided to sleep in a dilapidated canoe out in the field behind his barn. Sleep did not come easily for two boys nestled in their sleeping bags in the canoe. The stars were sparkling as we lay awake before sleep finally overtook us. . . . and they were still sparkling when we woke up. We always got an early start.

Our fishing poles were held tightly across our handlebars as we launched ourselves up the hill from his house on High Range Road, kicking up the dirt surface as we sped away. Our momentum improved considerably when we hit the short section of tar on Kimball Road. Then we pitched down the long hill by Faucher's farm before braking slightly in the still total darkness to make the curve on to our final stretch, Watts Road, another dirt road.

As usual, we were the first to arrive, but the broken call of a robin's song in the darkness forecast the coming light. Life was good, and we knew it! By eight or nine, as the frost was driven from the shadows by the sun's rays, we had been to all our favorite holes and were wet to our waists. Usually our next move was to dig out the Boy Scout frying pan that had made the trip in our packs. We fried up a half dozen trout as we warmed ourselves. We huddled around the fire, and as the steam rose from our wet clothes, it carried its own distinct scent to mix with the aroma of the sizzling trout. Oh yeah, I also burned Rick's sneakers a bit. Sorry, Rick!

*First published in April 1998.*

# Ivory Floats

As you while away this endless winter do you picture yourself in one of those TNN bass boat commercials? In a bass boat at full throttle leaping ahead of the others, screaming to the best fishing hole, as water sprays everywhere. *Not me*. I'm an *Ivory* man. I'm not talking about the softness of my skin or a nice relaxing soak in the tub. I'm talking about a wonderful boat about the size of a bathtub. Ivory is the name of my fishing buddy's jon boat. We call her Ivory because like the bar of soap, she floats! Ivory has provided thousands of hours of just plain relaxing fishing, usually on ponds without the sight of another boat or fisherman all day. Count on Ivory to take you away...

Ivory is small, in fact it is very small, only eight feet long. But what she lacks in length she makes up for in width, as she is nearly four feet wide, and very stable for a little craft. Ivory got her name over two decades ago probably because she kept us afloat in situations we had no right to stay afloat in. Especially considering her size and the amount of fishing gear we have managed to accumulate over the years. (You can never have too much "fishing stuff.")

One thing that is so nice about Ivory is that she can go anywhere. She fits nicely into the back of a pickup, with the tailgate shut. Slide her in a four-by-four and the possibilities are endless. Ivory is one fishing package! You may be thinking of some faraway mountain pond with only a ragged jeep trail to reach it. My definition of a remote pond is the one most fishermen only have a remote chance of fishing. You probably drive right by one or more of these ponds and lakes on your way to work or shopping, or at least when you are on that hour or more drive to the lake with a public access. Though New Hampshire has thousands of lakes and ponds, many of them are inaccessible because of the lack of public access. (Local camp owners and lake associations have closed many of the lakes in this state to public access.) If you are real lucky you can pay a ten- or twenty-dollar fee every time you launch your boat on one of the state's larger lakes.

A few years ago, my fishing partner and I took his big boat to a lake that had an obvious launch at a campground. We expected to pay a launching fee. But the fee was a year's membership at the campground! Well, we found a little dirt side road at this lake a couple of weeks later and slid Ivory right in. The fishing was fantastic! What I have found is that many of these "private" public bodies of water are generally underfished. Jet skis, pontoon boats, and high-powered boats line the shores adjacent to this expensive real estate. Most of the landowners are not the fishing types. I have caught lots of big old bass right under their yachts.

Generally, when we asked one of the shore owners if we could lug Ivory across their well-manicured lawn to put her in the lake, most have granted us permission. What an amazing sight it must be to see us lug this little bathtub-size boat to the water and gently set her in. Once we're on the lake, out pops the fishing gear. Like opening an umbrella, fishing poles and gear spring into action. Fishing poles stick out everywhere, enough to make a porcupine envious! After all we are bass fishermen! By actual count (I keep a diary) we have regularly caught over fifty bass a day on some of these "remote" ponds, right in your backyard.

Salmon fishing in Ivory usually means heading north, just after the April 1 opening date, to Lake Winnipesaukee, before the ice is out. We have had good luck trolling a smelt or a spinner in the open channels before the lake is ice free. Here again this little boat lets you access seemingly inaccessible waters. Frequently we have skidded her across ice to get to open water. The little thirty-five-pound thrust electric motor pushes her right along and a well-charged battery will last for hours. We have caught hundreds of salmon in this little boat by fishing the coves and watching the wind even after ice out. Lakers and nice rainbows can be found in the shallow water in the early season as well.

A small boat like Ivory can fit almost anyone's budget. Rick bought Ivory at a yard sale for fifty dollars. Although that was twenty-five years ago I'm sure there are similar bargains out there. Except for the electric motor, battery, and life jackets, you probably have everything needed for fishing the remote ponds in your area.

So stop wishing for horsepower and beating the crowds at the public launch sites. Relax . . . enjoy your fishing, become an Ivory man.

*First published in 1998.*

# Weird Woods Happenings

I've been an "outdoor" person day and night for all my life. I took to wandering the woods, even as a young child of six or seven, in far northern Maine near the Canadian border exploring the local stream and rivers for trout. I've never been afraid of the woods, even those times I had to stop and think a while of just which direction was out.

This includes nighttime walks and hikes most often not using a light, though one would be in my pocket. I have lived in Epsom for thirty-six of my sixty-five years and have learned the woods, fields, and especially the ten-thousand-acre Bear Brook State Park close by. Many times, my best friend of fifty years, Rick, who lives a mile away, has joined me. He too likes a nice nighttime hike.

It is on a few of these hikes that we have witnessed weird woods happenings. While I'm sure I could come up with a few weird wildlife stories it is the human ones that have on occasion had my neck hair bristling a bit.

For instance, one summer's night Rick and I were walking through the fields and woods that mostly separate our houses. We decided to cut through a stand of thick hemlock heading up hill towards a spot that lies a couple hundred yards off the dirt town road linked by a narrow tree-covered woods road. There is a small clearing here with a fire pit that has had a camper or two over the years. Not a spot most people would find and maybe a great spot to hide out.

As we walked toward the clearing, we could see light and realized someone was camping there. Sure enough, here was a tent and picnic table with a candle flickering in the middle. But no one seemed to be present. There never had been a picnic table here before.

Now for the weird part. Affixed to the big pine tree at the edge of the opening was a big plastic skeleton with a dagger driven through it, stapling it to the tree. Oh no, that was not the weird part. No; it was the tabletop scattered with electronic parts. Wires and small things were covering the tabletop. Something really weird was going on here and we were at a loss to what it might be.

In the back of my mind at this same time was the fact that there had been a number of bomb threats to state buildings in Concord. So I kind of wondered if there might be a connection. I figured I'd call someone the next day to report it. And as it turned out I didn't need to. The next morning when I arrived at my Fish and Game Region 3 office in Durham, to my surprise, there was the former head of the State Police Bomb Squad, who had recently retired from State Police and was now a deputy conservation officer. I explained what I had seen and he put my mind at ease when he said, "It's probably just some weirdo and don't worry about it." Easy for him to say, as this weirdo wasn't living down the road from him.

Another time, Rick and I were walking not too far from the last weird place on a moonlit night, cutting in off the town road into a sand pit. Here again it was a flash of light off to my left in a thick pine stand that caught my attention. It is amazing how, if you walk at night without a light for an hour or so, how light-hungry your eyes get. I mentioned to Rick that I had just seen a flash of light in the woods so we headed into the thicket to have a look.

Oh yes, another weird person. This person had hacked a narrow road off the pit road and up a slope into the thickest of the young pine. When we approached, there sat a big old Buick squat to the ground with trees covering the hood and top. So, we knocked on the window to see just who our new neighbor was. The back door opened and out slipped an old man with a cigarette hanging from his mouth. It was the lighter that gave him away. He was friendly enough, as best we could tell from his mumbling speech. The back seat he had exited was heaped with clothing and debris, looking like he pretty much lived there. But it was the front seat that was most interesting as there, taking up most of the room in this pretty big car, was a huge transmission. It was propped against the front seat like a bulbous passenger, leaving not much room for a driver.

It was his speech that was the weirdest. He claimed that President Bush, at that time the only Bush president, and even then it was a few years before, had agents looking for him and that he was in hiding. No, you just can't make this stuff up. But we were satisfied that he was no threat to us or our neighborhood and we walked on to leave him in peace. About a half hour later we heard his car start up and leave. His hide was up.

So a leisurely night hike in your neighborhood could be filled with weird woods happenings too. It's always a good idea to take a friend along on a night hike.

*First published in July 2015.*

# New Hampshire's "Rodney Dangerfield" Animal: The Fisher

Historically speaking, New Hampshire's fisher, like Rodney Dangerfield, got no respect. Although fisher are now widespread and abundant, many of the state's citizens would prefer to have the fisher exterminated or nearly so. Fisher are blamed for all manner of problems. The fisher has seemingly always been a creature of mystique, mystery and fear. Commonly called a "Fisher Cat," with the emphasis on cat, the fisher has endured the scorn of generations. Rabbits, partridge, pheasants, turkeys, horses, and children are supposedly all vacuumed up from the forest floor by the fishers' relentless marauding. This is *not* the fisher I know.

Fisher, along with most of the other furbearers, were nearly extirpated from New Hampshire by the time of the Civil War because of the unregulated trapping that had existed since the mid-1600s. Habitat loss also greatly contributed to the decline in fisher. The forested habitat preferred by the fisher was at a premium due to the agricultural practices of the times. Some measure of protection was afforded the fisher after the war, but it was not until 1934 that total protection was finally given to the few residual fisher left in the North Country.

The remnant population was able to rebound and spread north to south over the state themselves. Contrary to popular belief the fisher were not brought to the state from some outside source to control porcupine numbers; they came back once given complete protection. The fisher was again abundant enough in 1962 to declare an open season. A total of 126 were taken and soon the take grew to nearly a thousand by 1969, as a pelt brought over twenty dollars with a several-month season extending through the winter. By the early 1970s the fisher take topped a thousand at 1,130 in 1971 and 1,008 in 1972, by which time the pelt had doubled in value. By 1976 the pelt was worth a hundred dollars! But the population had crashed again due to too much trapping, and despite the lengthy season, only 277 fisher were trapped. The season was closed the next two years, then reopened in 1979. Pelt values had soared to over $130 and a shortened season brought another high take of 731 fisher.

The shortened season the next year also saw a season bag limit of one fisher imposed. The fisher pelt value kept right on climbing to $167 in 1986, but the restricted bag limits of only one, then three, allowed the fisher to once again return in numbers. As fisher numbers improved and pelt values declined the bag limit was raised to five fisher for several years. Two years ago the bag limit was raised to ten, except for five in Belknap and Carroll Counties. This brought a record fisher take of 1,187 despite a pelt value of under twenty dollars. During the December 1998 season with similar bag limits about two hundred trappers took 929 fisher.

Very likely such a high fisher take will again take the top off the fisher population and cause a reduction in numbers. Fisher, especially in the southern third of the state, are at high numbers based on an abundance survey reported by the state's trappers. Also, the trappers' success rate, based on the number of fisher captured during a particular time, indicates that fisher are very abundant across much of the state. The Fish and Game Department will be very closely monitoring any changes in population levels the next two years to forestall another collapse due to overtrapping. We do not want history to repeat itself!

Four different scientific studies of the state's fisher show that fisher prefer small prey species, namely mouse-sized mammals that contribute about a third of their diets. Biologists have looked at the stomach contents of fisher in 1968, 1977, 1980, and 1987. All produced similar results. Mice, small birds, fruit and berries, and deer in the form of carrion made up most of what a fisher eats. Snowshoe hare made up less than ten percent. Contrary to popular belief, housecats are not a regular food item. Cat hairs were found in only one of the over one thousand stomachs examined in 1979 and 1980.

Fisher remain one of the most secretive and mystical characters on the wildlife scene. They will continue to make New Hampshire a very special and wild place to live.

*The last two decades have seen a dramatic decrease in fisher population. In the 1970s, over one thousand were trapped every year, down to only thirty-six in 2022. A new type of rodenticide might be a contributing factor.*

*First published in 1999.*

# How Much Wood Can a Wood Chuck Chuck???

How many bear, moose, ducks, or turkeys are there in New Hampshire? Or how many bass live in my lake? Are the fishermen taking too many? Does my lake need to be stocked with more fish? These questions all deserve an answer. Hopefully a well-founded answer. But like the question begging an answer about a woodchuck's ability, none of them come easily. Why? Fish and wildlife are very complex and dynamic resources. Constantly changed by births, deaths, migration, immigration, and seasonal changes.

Fisheries and wildlife biologists have tried to measure these changes for decades. I liken the complex wildlife populations to a constantly changing puzzle. We can learn pieces of the puzzle a bit at a time. Then we try to factor them into the whole puzzle picture. A picture that is constantly changing. A difficult task to say the least. Yet a lot is known about our fisheries and wildlife populations. For some species, we have had the abilities to develop sensitive population indices. But for most species, it is far more important to gauge the general population level over time rather than an exact count at any given moment. It is more important to know whether a population is increasing, decreasing, or relatively constant. And we need to know what factors are causing the changes and whether we can influence the changes. So how do biologists measure fish and wildlife population changes?

For example, waterfowl such as black ducks, mallards, geese, and sea ducks are observed annually by at least seven different methods to assess their numbers. And since waterfowl are migratory, these same observations are made simultaneously up and down the Atlantic coast from the Canadian Maritime provinces on down the US coast. Data from the waterfowl observation are provided to the US Fish and Wildlife Service.

A longstanding survey conducted on the East Coast Atlantic Flyway has been the midwinter aerial survey. In New Hampshire, this involves a several-hour small airplane ride at low altitude around Great Bay, out to the Isles of Shoals, then back south along the shoreline to Seabrook, and finishes up with several transacts of the Hampton Marshes. This survey has been conducted since the early 1950s. Each year over five thousand different ducks and geese are spotted this way. I have been accompanying our waterfowl biologist Ed Robinson on this flight for nine years. Usually it occurs the first week of January all along the flyway. It is amazing to see how different it can be from year to year. Some years the bay is practically entirely frozen and some years mostly open, like this year.

For about eight years we also have been doing a winter inland census. This consists of the regional biological staff examining open waters, where mallards and black ducks are wintering. For me in Region 3, it means checking for "city ducks." I recently counted over a thousand ducks, mostly mallards, wintering in and around Nashua. We are always looking for these winter concentrations so if you know one, give your local Fish and Game Office a call. In January 1996, the state's total included 5,592 mallards, 357 black ducks, 195 Canada geese. Plus 8 common mergansers, 5 hooded mergansers, 12 mute swans, 2 greenwing teal, 72 common goldeneyes, 2 buffle heads, 1 pintail, and 16 mallard/black duck crosses were counted. Each April, a waterfowl nesting survey is conducted. This involves biologists observing each of sixty plots scattered across the state to count nesting ducks and geese. This survey has been done for about ten years. Each plot is about a thousand yards on a side. I have four sites that I visit in Region 3 and an assistant does another six in Region 3. The biggest change I have noted over the last few years is the big increase in nesting geese.

For the last several years, a group of volunteers has been cooperating with the Fish and Game Department to monitor waterfowl on Great Bay over nearly the entire winter. On selected weekend days volunteers are stationed at various prominent points around the bay and simultaneously count ducks. By counting over a short period at the same time, duplicate counts are minimized.

In addition to the state counts, the US Fish and Wildlife Services also monitors duck hunters who have purchased duck stamps. When I purchased my duck stamp this year, I was given a card to keep track of each day I hunted. What species of ducks taken was asked as well as were any cripples lost. Then at the end of the season, they sent me another survey card so that I could mail them the data. Other hunters are asked to collect a wing from each duck they shoot. The wings are collected by the Service and sexed and aged by species. This way, the impact of hunting on the duck populations can be measured. It also provides biologists with the success of the previous breeding season by calculating how many juvenile versus adult ducks were harvested.

Lastly, each state is given a quota by the Fish and Wildlife Survey of ducks and geese to be banded each year. In New Hampshire about six hundred to eight hundred geese and over a thousand ducks are captured and banded. Female wood ducks and hooded mergansers are captured and banded in nest boxes in May. By late June, the goose roundup begins. Geese molt and lose their flight feathers at this time. So we simply build a corral and drive them into an enclosed pen. Preseason duck banding is done in late summer at several locations around the state. This is followed by mid-winter banding in January or February. Band reports from these efforts, primarily reports from hunters in the fall, also help determine the impact of hunting on the duck populations. Over the last several decades, the waterfowl information collected has helped all the states provide the best management possible. In New Hampshire, the duck season was adjusted by three or four days from October 1 to the 3rd or 4th, because of banding hen wood ducks in nest boxes. Waterfowl seasons and bag limits are adjusted each year based on the analysis of the accumulated data.

As you can see, counting a species is never an easy task. The Fish and Game Department is continually assessing our natural resources from animals as small as a shrimp—yes, our marine biologists do shrimp counts—to our mega moose. Will we continue to keep track of the fish and wildlife? You can count on it.

*The New Hampshire Fish and Game Department continues to band and survey waterfowl each year. This is an ongoing effort.*

*First published in 1996.*

# New Hampshire's Mid-Winter Coastal Waterfowl Survey Takes Flight

Saturday, January 15, 2005, 8:26 a.m.: The right wheel of the gold Cessna 172 spun to a halt seconds after takeoff from Concord Airport. We were finally into the air for the annual New Hampshire mid-winter coastal waterfowl survey. Thank goodness we opted for the next flyable day, which happened to be a Saturday. Fish and Game staff have conducted the annual mid-winter coastal waterfowl survey since 1952. It normally took place the first week of January each year. But this year, we were delayed four times due to poor flying conditions.

This same survey is conducted all down the Atlantic coast from the Canadian Maritime provinces to the mid-Atlantic coast. That way all the ducks and geese are counted at the same time before much movement can take place. What a glorious day for flying. Pilot Dick Meyers from Concord Aviation was at the controls of the Cessna. I was beginning to think he was too timid of a pilot since he had postponed the flight so many times.

Immediately I noticed that we were taking off on the snow/no-snow line in New Hampshire. Off towards the north and west, the ground, including the fields, looked pretty well covered with snow. But below us, as we turned eastward to follow along Route 4, the fields were essentially bare. The heavy rain events of Thursday night into Friday set winter back once more.

We approached Great Bay from the southwest as usual per my directions to the pilot. I have flown the annual winter waterfowl survey since about 1990. Since 1994, Julie Robinson has been the other Fish and Game observer. Julie concentrates on counting the waterfowl. She called the numbers into my ear via the headsets we were all wearing. I recorded the numbers on data sheets: one for Great Bay, another for the Isles of Shoals, one for the coastline and one for the Hampton Marshes. I also navigated the trip. I made sure we covered all open water, but did not duplicate our counts. I had to constantly keep track in my head of where we were, where we had been, and which flocks of ducks and geese had been counted from Julie's perspective, all the while as the plane dipped and turned round and round, and as I wrote the numbers down.

As we approached the bay it was obvious that there was practically no ice. All the rivers converging on the bay were free of ice as well. I immediately determined our strategy for this flight. We turned south, dropping altitude as we headed up the Squamscott River all the way to Exeter, as the river was all open save for a patch of ice with a gap in the center by Swasey Park in town. The sewer lagoons next to Route 101 typically hold some numbers of ducks. We retraced our flight back to the south end of the bay.

Now we began a counterclockwise trip around Great Bay itself. Since Julie was in the front right seat she needed to scan right. I coached the pilot down and just far enough off the shore to count the bunches of geese tucked along the shoreline. Most years it is a battle to convince the pilot to fly as low and slow as they dare. Usually we try to get down to two hundred and fifty feet or so.

Julie began to fire numbers at me: "Twenty geese, five geese, thirty geese, eight black ducks, fifty geese; 'closer to the shore,' twelve geese, and holy smokes a hundred geese." The plane dipped and turned sometimes gut-wrenching turns, to stay over water and away from the houses as we skimmed the mocha-colored water while the tide was receding.

The pilot called the Pease Airport tower, as is protocol, to alert them and to seek permission to fly near the base at low altitude to conduct this required federal waterfowl survey. He called to them again to alert them to our location as we approached the end of a runway. In our headphones we could hear the control tower call to other planes, landing or taking off from the runway we were about to cross several times, as we swooped over the bay. We still needed to obey the safety rules to keep the public from the risk of such a low-flying aircraft. So, tight turns were in order whenever we flew over land or up the open rivers.

My concern for the pilot's timidity quickly vanished as we dropped down to about two hundred feet over the stained water. "We've dropped to one hundred and fifty feet here now," Dick echoed into my ears. At that point, even I could see blades of grass along the shoreline. The ground seemed awfully close! Immediately, I suggested two hundred feet would be fine with me.

It took three complete revolutions of Great Bay to close the gaps to be sure we had covered the expanse of open water. Dick pulled back quickly on the yoke and applied power to jerk the plane to the required one thousand feet as we left the vastness of the bay on our side trips up the Lamprey and Oyster Rivers. Geese were just about everywhere! Scattered clumps of ten, twenty, six, fifty, and soon three rows of numbers lay on the clipboard on my lap.

We then crossed over to Little Bay for a once around, then up the Bellamy River and down the Cocheco River. Minutes flew by as well. Then we headed over to the Salmon Falls to the head of tide and down river, scanning the river with Julie calling off more and more goose numbers. In all, 2,708 geese were counted in the Great Bay area. Last year, 2,665 were counted. Even though the geese were scattered all over the place in small groups, the total number was very similar to recent years' counts. Over time, the number of wintering geese has gradually increased in the bay. In the 1950s, it averaged 1,207, in the sixties, it was 1,692 and by the eighties, the average was 2,171. The average increased to 2,235 by the nineties and since 2000, it has averaged 2,396 per year.

The trip down the Piscataqua River was less eventful as we had to gain altitude once more for our trip over the city. I called to the pilot for a glance at Back Channel to check for the mute swans that were supposed to be wintering there. Then, the pilot pulled back on the yoke hard and gave the Cessna full throttle to climb to thirty-five hundred feet before we peeled hard further east and headed out to the Isles of Shoals. He reminded us that he wanted some gliding room before heading out across the expanse of ocean that lay between the Shoals and us. It seemed so close at first glance, but the huge lobster boat trailing a froth-covered wake looked small as we passed it at thirty-five hundred feet. We again spiraled down to four hundred feet at the isles and began a methodical search of each rocky foam-covered island jutting from the cold steel gray sea.

Here we found numbers of eider ducks dodging the swells as they braked over the granite fingers clasping the sea. Males are huge with white speckles, while the brown females are hardly discernable. By our count, eighty seals lay basking on the northern ledges on the lee side of a tiny rocky island. Even though this probably is the riskiest part of the flight, way out at sea, it still was my favorite part of the annual survey. Despite the presence of buildings on the islands, they seemed empty, stranded at sea like some abandoned ship. Waves rolled against the rocks, shooting white foam onto the multicolored granite ledges, with my mind taking slow-motion snapshots of the scene that lay below, almost mesmerizing me. It was a scene of an angry-looking turmoil yet with complete tranquility from my perch. We counted 884 eiders and 36 goldeneyes while weaving among the islands.

As we headed back from the Shoals, we spiraled back up to thirty-five hundred feet. Preferring, I supposed, that if the engine failed we would crash-land on some rocky granite patch rather than be swallowed by the even more unforgiving cold sea.

New Hampshire's short coastline trails off rather quickly from the air with only a few diving ducks noted. Scoters are hard to see, even at low altitude, as they tend to dive underwater for extended periods, making them a low-count duck, despite seeing numbers from a shore view. We did count a few goldeneyes too. We swung in over several of the salt marshes, which were completely devoid of ice and snow. I can't remember a year since I started the survey that has been as free of ice and snow as this one.

The Hampton Marshes provided a low count despite their open condition and a low tide that should have revealed numerous black ducks, as it has in years past. They simply were not here this Saturday. I think there was so much open water in the rivers and even bare fields that the ducks were scattered this year and not concentrated along the coast as is normal for early January. The goose counts seemed normal but the mallard and black duck count was low by my recollection.

As we gained altitude back up to twenty-five hundred feet for our trip home I slumped back into the rear seat as the intensity of the morning gave way to a relaxing cruise. The Pawtuckaway Mountains, the remains of a long-dormant volcano, framed my view to the north, then Fort

Mountain, in my town of Epsom, protruded starkly against the winter sky as we then glided over the huge expanse of Bear Brook State Park before pitching down, then landing gracefully back at the Concord Airport a little before 11 a.m.

Checking the 2005 numbers against past surveys showed some interesting trends. First it should be noted that the New Hampshire count is just a small part of a coastwide count. The Maine team flies for five days to cover all the islands and coves to complete their count. Although we can compare the New Hampshire figures from year to year, it really is the total Atlantic-wide count that is significant. After all, some years, the ducks and geese have migrated past this state in a cold icy winter and are simply counted somewhere else. For instance, the mallard count at 251 was similar to recent years but the black duck count of only 234 was much lower than the usual 800 counted. In 2002, 2,292 black ducks were counted.

Long-term New Hampshire trends show a growing number of geese, mallard, and eiders wintering along our coast. The black duck count has been pretty stable over time. New Hampshire's coast, especially Great Bay, is a major wintering area for the region for a number of species including bald eagles. I am always looking forward to next year's survey.

*As of 2024, New Hampshire Fish and Game is no longer doing the aerial surveys. US Fish and Wildlife is now doing these surveys.*

*First published in January 2005.*

# I've Got Friends in Low Places

I've got friends in low places, where the mosquitoes abound and the deer chases the moose away. Actually, I've got lots of friends, or at least acquaintances, that hang around the low places. In New Hampshire, they are generally wood ducks, mallards, black ducks, hooded mergansers, and Canadian geese. Occasionally a stranger such as a pintail passes through.

The New Hampshire Fish and Game Department's waterfowl biologist, Ed Robinson, initiated a renewed effort in 1987 to band ducks in the Granite State. Our efforts are part of a US Fish and Wildlife Service Atlantic Flyaway banding program. We have been very successful! A total of 4,710 ducks have been banded and since 1991, we have added 3,301 resident Canada geese. I have met most of these birds. Since I am the regional wildlife biologist in Region 3, the southeast corner of the state, I happen to work in low places such as marshes and swamps where many of the ducks and geese breed or over-winter.

Duck banding is done at three specific times and places. In mid- to late May, regional biologists and technicians fan out across the state wildlife management areas containing duck nesting boxes. May is when the female duck is midway or more through the twenty-eight-day incubation period of her ten to fourteen eggs. We quietly approach each box, carefully, very carefully, stand up in our canoe, and stuff a rag into the duck box opening. We trap the hen inside the box and carefully open the side panel to see what is inside. I find it exciting to see who's home. Both wood ducks and hooded mergansers are cavity nesters and are banded during this period. Sometimes I've been surprised by a big hornet's nest! Have you ever tried to do anything rather quickly while standing up in a canoe on a windy day, hugging a tree and flailing away with a hammer in your free arm?

This job is challenging.

Preseason summer banding occurs in late summer just before the fall duck season begins. Will Staats, regional wildlife biologist, has perfected a method to get dozens of wood ducks under his rocket net along the upper Connecticut River. Mallards are targeted in the southern counties.

Winter duck banding has been done near the coast in Dover and North Hampton for nearly a decade. All the geese are banded in mid- to late June when they lose their flight feathers for a few days. The goose roundup crew travels around the more urbanized areas where the resident geese breed. Golf courses, office parks, condominium ponds, and a few farm ponds are visited each year. We simply build a covered pen with snow fence, create long wings on each side with netting and walk the geese right into it . . . usually.

All the leg banding information is sent to the US Fish and Wildlife Service bird banding laboratory in Maryland, where it is computerized. Decades of banding information are available.

So, how do we know where our ducks and geese go? Well, we keep track of our acquaintances just like you. I'll bet if you read a daily paper like I do, you tend to glance at the obituaries to see if you recognize anyone. It also is nice to know everybody listed is older than you. Hopefully, way older! Well, the bird lab maintains the duck obituaries. Most ducks and geese that are reported come from hunter takes. In fact, if you take a banded duck or goose, please report it. In a few weeks you will receive an official death certificate for the duck or goose you got, providing you with a history of where it was banded. Some are bizarre.

From 1992 to 1997, a total of 209 New Hampshire banded ducks have been reported. This included 112 spring/summer banded birds and 97 winter banded. About 43 percent of the banded ducks were recovered in New Hampshire. The majority, 57 percent, headed elsewhere before hunters took them. Spring and summer banded ducks were generally taken south of New England, some as far south as Florida, but some ducks head west to Missouri, Michigan, Minnesota, and beyond. The winter-banded ducks tend to head north from New Hampshire. Our short eighteen-mile coast and the Great Bay Estuary is their winter "fun in the sun" place.

We had the mother of all winter parties one cold February morning in 1994. The thermometer hovered in the low teens; the snow was uncharacteristically deep, even along the coast. The gray cold Atlantic was churned up by a cold east wind just a half-mile east of our banding site. The roar of the surf could be heard in the distance. The snow and cold conditions had piled up the ducks this winter day at our banding site in North Hampton. The roar of the rockets flinging the giant net over the ducks always brings an adrenaline rush. The sight we held as Ed Robinson and I sprinted to the net a hundred yards away spiked our adrenaline flow even more. The net was full of ducks! More ducks than we had ever seen before. In fact we had captured over six hundred and fifty, mostly mallards, but nearly two hundred black ducks. Black ducks are the wariest and most difficult to catch.

Over ten of the ducks we caught that day have been reported. Many of them went north of Maine, Quebec, and Ontario, and one to Newfoundland. But some were killed south of New Hampshire in Massachusetts, Delaware, and Virginia.

Since 1991 we have banded 3,301 resident Canada geese. The majority, 65 percent (265) of the 409 reported killed, have been taken right in New Hampshire. These truly are "resident" geese. About 89 percent of the reports have come from New England. During a severe winter the resident birds tend to go south into Massachusetts or New York. But five were recovered to the west in the Mississippi Flyway, and thirty-two were recovered in the mid-Atlantic region of the Atlantic Flyway.

The little wood duck seems to be the most traveled duck. However, it's clear to me what an odd duck is: it is one who traveled for some unknown reason to an odd place. Or, maybe New Hampshire was the odd place, and he was really from "there." Unraveling the mysteries of our waterfowl will continue to be an exciting quest. Many surprises lie ahead, even an odd duck or two.

*It varies from year to year, but every June, between six hundred and eight hundred resident Canada geese are still banded as of 2024.*

*First published in 1998.*

# How Old Is Old in Animals?

Hey, you old bat! Well, how old is an old bat? Or for that matter, how old do deer, moose, bear, fish, ducks, and turtles get? The answers will surprise you.

Most animals do not survive very long in the wild. Often mortality rates, especially for the young born each year, are upwards of 50 to 60 percent or even higher. In fish, often fewer than 1 or 2 percent of the eggs that hatch will live to be adults. About half of the bear cubs born each year do not survive to be a year old. Diseases, predators, deep snow, starvation, drought, people, hunters, and cars take their toll.

So how old do they get? We have been collecting aging bear teeth since 1978 in New Hampshire. So far, the oldest in the wild was a thirty-year-old bear killed in 1992. We also have had bears aged at twenty-seven in 1978, twenty-five in 1993, and twenty-six in 1995. The average age of bears in 1995 was only 4.5 years for males and 7.11 for females.

You would expect to see a lot of old moose wandering the state. After all, they were protected from hunting from 1908 to 1988. Surprisingly few old moose exist. Since 1988 we have seen aging moose taken by hunters. Moose teeth wear out quickly and by age ten or twelve they are nearly worn down to the gums. The oldest moose aged was fourteen and a half. Moose do regularly get to be eight years old, but the average age was three and a half in 1995.

Deer are short-lived creatures. Fawns are regularly eaten by coyotes and bear, based on studies done in other states. In fact, in downeast Maine, deer fawns made up 75 percent of the diet of coyote pups being weaned. We do not know what percentage occurs in New Hampshire, but likely it is considerable. Hunters certainly are the controlling factor in the age of most adult deer. Each year, biologists examine several hundred deer at the biological check stations during the first five days of the firearms season.

In 1996, I examined 132 at a biological check station in Manchester. Deer are aged by examining their teeth. Deer get new teeth only until one and a half years old, then gradually wear their teeth down to the gum line over the next five to seven years. Deer teeth therefore are almost completely worn beyond use by the age of seven or eight. Therefore, deer cannot get very old in the wild. Of the 132 deer I aged, 81 percent were killed by the age of two and a half, and 93 percent by the age of three and a half. Only 3 percent of the deer were aged at five and a half and six and a half years.

We have collected hundreds of teeth from coyotes, fishers, and otters for aging as well. Very few of these are ever over five years old. In 1993, one fisher was age eight and one age nine.

Fish have the highest mortality rates, but fish can produce thousands or even over a hundred thousand eggs to compensate for this loss. Some fish, like bass and lake trout, can outlive most mammals. Lake trout have been aged at eighteen years at least based on annual rings in their scales. By the way, an eighty-pound lake trout caught in Great Bear Lake in Manitoba, Canada, was one hundred and fifty years old. Bass frequently make it to the teenage years. The oldest aged in New Hampshire was seventeen years. From studies done in New Hampshire, we know that a smallmouth bass that is seventeen to eighteen inches long is eight to ten years old.

The results of duck banding have given us records of ducks living to at least eleven years old. In a banding effort in Dover in 1994, two ducks were captured that had worn-looking bands. One mallard had been banded eleven years before in New York and the other mallard seven years before.

Turtles can be aged by counting the rings on the plates on their belly. Turtles regularly live to be sixty to eighty years old.

Now, about old bats. While doing winter census of bats in caves in the winter of 1985, a single bat was seen that had a band on it. An inquiry found that the bat was indeed an old bat. It had been banded in that very cave twenty-seven years before. Incidentally, a check of this cave in 1995 found this same bat still there. He is at least thirty-seven years old.

*First published in 1996.*

# Fish and Wildlife Restoration in New Hampshire: A Century of Successes!

What a difference a century has made in the abundance of fish and wildlife in New Hampshire. The wildlife we see and encounter practically every day of our lives was unthinkable just a few decades ago. In 1900, wildlife was encountered only in stories told by the elders. The deer population was decimated by the time of the Civil War; moose had vanished to all but a handful in Coos County; the last wild turkey was eaten by a farm family in 1853. Bears had been killed off too. The few that remained had a ten-dollar bounty on their heads. Beaver, otter, ducks, and geese were all slaughtered in untold numbers by the settlers without regard to seasons or limits, or perhaps even awareness of what the culture of the times was doing to nearly all wildlife populations. By the turn of this century most wildlife was gone or on the verge of extinction.

Concerns had been voiced for the lack of laws or protection of wildlife. In fact, the first law protecting any wildlife species was passed in 1740 to set a deer season "for the entire Province" from August 1 to December 31. However, despite the passage of various laws over the next one hundred and fifty years from that first law, there were virtually no means to enforce them. Most laws remained ignored. Dams went up along our rivers that blocked the migration of fish to spawning grounds and the rivers were turned into open sewers for the most part. By 1900, fish and wildlife populations had been decimated as well as their habitats.

The New Hampshire Fish and Game Department was formed in the latter part of the 1800s specifically for the purpose of returning salmon to the Merrimack River. The department met with limited success, but funding was dependent on an appropriation by the legislature. Fish and wildlife was generally not of much concern to the legislators or most of the citizens either, so funding was never very generous. In 1891, the legislators appropriated six hundred dollars to hire detectives to prevent the killing of deer in the deep snow in Carroll and Coos Counties. The following year only two hundred and fifty dollars was appropriated.

Finally in 1903, the legislators required the first deer hunting license for nonresidents, at ten dollars. A total of 247 were sold! By 1905, nonresidents had to buy a license for all game hunted between October and December. The income jumped to ten thousand dollars.

By 1909, a resident hunting license was required for the cost of one dollar. Revenue soared to $26,848 in 1909, but the legislators had only budgeted the department for $8,600, so the balance went to the general fund. By 1913, all license revenue was earmarked for the department. Since that year the department has relied exclusively on the revenue generated by license sales. Revenue was boosted dramatically by the requirement of a one-dollar fishing license in 1917. By 1930, the year a trapping license was required, revenue topped the fifty-thousand-dollar mark. The department's fund was further bolstered by the passage of the Pittman-Robertson Federal Aid to Wildlife Act of 1938, which collected an 11 percent tax on arms and ammunition that was reallocated to each state. The Dingell-Johnson Act followed some years later, providing funding for fisheries work by a federal tax on fishing equipment.

Funding by the sportspeople of the state, coupled with a shift in the culture, led by the same concerned sportspeople, has brought nearly a complete restoration in the fish and wildlife populations compared to the beginning of the 1900s.

The deer population has recovered from as few as five thousand at the turn of the twentieth century to about eighty-five thousand now. There were believed to be about a dozen moose in Coos County at the turn of the twentieth century and the current population is estimated to be nearly ten thousand. Moose now occur in nearly every town in the state. The ten-dollar bounty on bear was increased to twenty dollars in 1947. Finally even this bounty was repealed in 1955 and bear were protected with a season in the early 1960s. From as few as three hundred to four hundred bear in the earlier part of the twentieth century, we now have about four thousand bear inhabiting all but the most urban areas of the state.

Turkeys required more than just protection because they were extinct by 1855. A successful reintroduction effort in 1975 was begun when twenty-five wild turkeys were captured in New York and released in New Hampshire along the Connecticut River. And succeed they have! We now have twelve thousand to fourteen thousand turkeys in the state. Beaver had been trapped out by the 1880s but with protection beginning in 1901 they have made a full recovery as well. For the most part the beaver returned on their own except for a few moved from Coos County in the 1930s, which were released into the central part of the state. The return of the beaver and the thousands and thousands of wetlands they have restored has bolstered the return of mink, otter, wood ducks, and scores of other animals who live in the wetlands created by the beaver.

The Fish and Game department has a long history in the restoration of fish populations, beginning in the 1880s. The department has operated fish hatcheries since before the turn of the century that have contributed to the fish restoration. But few realize the significance of some projects the department has undertaken. In the 1960s and 1970s, the department constructed fish ladders on most of the rivers on our coast, which enabled the return of hundreds of thousands of river herring to bolster the tide-water ecosystem. The ladders were funded by sportsmen as well and often built by the department's own construction crew. As a result of the return of fish, the last several years ospreys have returned to nest at Great Bay. Osprey frequently feed their young herring restored by the construction of fish ladders. Although fishermen, hunters, and trappers have benefited tremendously from the restorations of game, fish, and wildlife populations, many non-hunted species have been restored as a result of habitat restoration and protection of the game species. Non-sportsmen, who have not contributed license fees toward the restoration efforts, have also enjoyed the abundant populations of many game and non-game species.

Hunters, fishermen, and trappers continue to generously fund the department. The department continues to be a self-funded department as envisioned in 1903. From a budget of about twenty-five hundred dollars in 1903, the department will finish this century with a budget of about seventeen million dollars. Over three million dollars of these funds are provided by the Federal Aid in Fish and Wildlife Restoration Programs.

As we go into the next century, we should all be grateful to those who had the vision and the dedication to right the terrible wrongs done by our forefathers. You have a greater chance of seeing a deer, moose, bear, or turkey out of your home or vehicle than did your great-grandmother. However, the sheer numbers of people now living in this state could have a dramatic effect on wildlife populations by the amount of habitat that is lost to development. It will take a lot of work and funding by hunters, fishermen, and trappers to change the culture of the general public to recognize the value of abundant fish and wildlife populations. The easy work (restoration) is behind us. Can we keep our wildlife with the ever-burgeoning demands of humans? If the past is a reflection of our future then we will succeed!

*First published in 2001.*

# Sowing the Seeds of Conservation That Will Last a Hundred Years

If you are a New Hampshire sportsperson, it's that time of year again to belly up to the local sporting goods counter to purchase next year's licenses. Surprise! You've got to pay a little more this year than last, on top of a significant across-the-board increase for 2002. Perhaps you too wonder, "Where does this money go?" "Gee, with all the troubles Fish and Game seems to be having lately, is my money really going towards the things I think it should?"

Well, let me tell you what the last quarter of a century as a wildlife biologist at Fish and Game has shown me over and over again. Nothing has ever deterred the staff from its heart-and-soul commitment to this state's fish and wildlife resources. Often, however, the women and men who buy

*Eric Orff planting an apple tree.*

fishing, hunting, and trapping licenses have not been recognized for their profound effects on the fish and wildlife populations. A turkey here, a moose or bear there, or a herd of deer grazing in a field are all commonly seen today. It seems like we take the wealth of wildlife in our lives for granted, especially by those who *do not* buy any license. Your purchase of licenses makes a difference. In fact, it makes *all* the difference! Without your support there would be no Fish and Game Department. Period! The Fish and Game Department has about a seventeen-million-dollar budget each year that comes right out of the pockets of the state's sportsmen and women.

The majority of the revenue comes directly from license sales, but several million dollars each year come to the department through the Sportfish and Wildlife Restoration Funds. These are monies collected by the federal government through special taxes on firearms, ammunition, fishing equipment, and boating-related taxes. Most of these were self-imposed by the nation's sportsmen as a way to further their conservation concerns. Sportfish and Wildlife funds are distributed to the states by the US Fish and Wildlife Service as reimbursement for money already spent on various federally approved fish and wildlife projects. Much of these funds are earmarked for research and management projects. Three quarters of my biweekly salary comes from these funds. The other quarter of my salary comes from your license fees.

Let me give you a few examples of how I have been spending your license dollars this year and even give you a little history of where your money has gone in the past.

In early January waterfowl biologist Ed Robinson arranged for the annual winter waterfowl flight. This means getting a plane and pilot from the Concord airport to fly me and my fellow Region 3 biologist Julie Robinson on a several-hour tour of Great Bay, the Isles of Shoals, the coastline, and Hampton Marshes to do a low-level count of the waterfowl. We always elicit a rolling of the eyes of the pilot when I stand in front of him as we ask him to fly "as low and slow" as possible over the nearly frozen waters so we can count ducks. I'm kind of a big guy and not many pilots seem happy to fly a three-hundred-pound gorilla low and slow over the wind-tossed

Atlantic Ocean in early January. But they do. This midwinter flight is done up and down the Atlantic coast each winter to gauge duck numbers and has been done for over fifty years thanks to your support.

We also do some midwinter duck banding and each regional biologist goes out on the ice to check duck nesting boxes over the winter to gauge their success the previous spring as well. There is a lot of waterfowl work that needs to be done each year in order to properly manage a multitude of species. Hunter's dollars make it happen.

There are grouse and turkey census routes that need to be run at sunrise each April, as well as woodcock census routes and waterfowl nesting plot surveys. Spring, like fall, is a very hectic time for wildlife staffers like me. We also conduct several prescribed burns of state lands each spring in cooperation with other state agencies in order to improve the wildlife habitat on state owned lands.

May, well, May produced the seed for this article. On May 6, 2002, I was planting a group of twenty crabapple trees on a hilltop field of a farm in Farmington when it struck me that what I was doing could very well last a hundred years. Certainly apple trees can live that long. I won't. I had been working with the owner of this hundred-acre farm for a couple of years. First, using funds from the habitat stamp that was started three or four years ago, we brought in a big machine to reclaim the old field that had been overtaken by small trees. Field land is at a premium in New Hampshire, since much has been lost to development or has simply been left to grow back to woods. Turkeys, cottontail rabbits, woodcock, and many other species need these open areas to mate, nest, and rear their young. It is expensive to bring in the brontosaurus to literally shred the trees and reclaim the fields. Fortunately, the habitat funds paid by each licensed hunter provides the monies needed. There is a conservation easement on this farm so I know that it will be kept as open land for over a hundred years.

I offered to plant twenty crabapple trees in one corner of the new field to add a winter treat for the local turkey population. These trees are purchased thanks to the initiative of turkey biologist Ted Walski. Turkey license fees are used to purchase over six hundred crabapple trees each year. Some that I planted the first year of this effort in 1997 are already laden with apples. These will be full-sized apple trees some day and should bear fruit for decades. What a wonderful feeling it is to know that you are doing something that will benefit wildlife for a hundred years.

New Hampshire's deer, moose, and bear populations have all expanded during the last couple of decades I have been working at the department. I have been intimately involved in the bear population for much of that time. The bear population increased from an estimated twelve hundred to fifteen hundred bears in the early 1980s to about five thousand now. Just this fall I had a conservation officer report watching a sow and cubs in Fremont one evening. Wow, bears well into Rockingham County; unheard of when I started at the department. Black bear had been gone for over a hundred years from much of the southern part of the state.

Along with an increase in bear numbers has come an equally important shift in the public's attitude about bears. Again, paid for by you the hunter. When I first started working on the bear project in the late 1970s, the New Hampshire Farm Bureau was seeking legislation to move the opening day of bear season from September 1 to August 1 to further reduce bear numbers. There was little public interest in bears. Sex and age data on bears collected in the late seventies and early eighties indicated a dwindling bear population. Hunters paid for getting this information. This information was just what was needed to convince the legislators to give the department the authority to regulate bear hunting starting in 1985. The declining bear population was quickly turned around and the population has grown in numbers and expanded its range ever since.

More bears over a wider area also meant that the department needed to begin teaching people how to live with the bears. By 1995, the department kicked off the "Something's Bruin In New Hampshire" campaign. This project was modeled after the successful Brake For Moose program. One of the goals was to get the state's residents accustomed to having more bears around. Thousands of bear bumper stickers were distributed as well as several informational sheets dealing with bird feeder/bear problems, backyard bears, and even trash and bears. The department had thousands of postcard-sized bear cards printed with distribution to campgrounds, rest areas, and any other place that a little education would help teach people to live with bears. Each winter I do an inventory of the "bear" supplies and gear up for the next summer. The bear license monies pay for these supplies. Hunters' fees have paid for a change in the very culture of the people who live in New Hampshire so that we can have more bears.

I know there will be bears in New Hampshire a hundred years from now thanks to the funding provided by New Hampshire hunters. Hunters, when you pay for your license you are making an investment in the future of your great-great-grandchildren's lives.

*First published in 2003.*

# Fish Climbing New Hampshire's Ladders of Success

Looking back fifty years or so at the fish and wildlife management programs of the New Hampshire Fish and Game Department, a couple of initiatives really stand out for their success. On the wildlife front, there is no doubt it was the very successful reintroduction of wild turkeys. Much has been written of this success. But down on New Hampshire's coast an equally, or more successful, program has gone practically unnoticed.

This coastal success story is the restoration of a relatively small fish, the river herring, or alewife. Though small in size, usually less than a foot, its impact has been in numbers. Each spring literally tens of thousands of adult herring ascend the coastal rivers to spawn, all thanks to the construction of fish ladders, or fishways as they are technically termed.

I have closely monitored the construction and operations of New Hampshire fishways for nearly forty years. In fact, in the early 1970s, while I was the chairman of the Merrimack River Watershed Council, our organization intervened in the relicensing of the Amoskeag Dam in Manchester by the New Hampshire Public Service Company. To that point the renewal of the hydro station license was a given during the periodic process. We requested of the Federal Energy Resources Commission that the license only be granted pending a plan to construct a fish passage facility at the dam. Since then I have helped the New Hampshire Fish and Game Marine staff occasionally at one of our coastal fishways, from counting alewives to removing a very angry stuck beaver.

Like nearly every river in the Northeast, New Hampshire's coastal rivers were used to generate power for the early mills established on them as early as the 1700s. Dams were built to harness the only available power of the times and literally fenced out the fish that had ascended the rivers for millennia to reproduce. Human growth along the coast further impacted these same rivers as they were essentially turned into open sewers, further decimating the residual fish populations. Herring, Atlantic salmon, lamprey eels, American shad, and even sturgeon, which could grow to lengths nearly ten feet long, were also eliminated from all these rivers. Basically, until just a few decades ago these rivers were nearly dead save for a few species like smelt that managed to survive in the tidal portions of these rivers.

All that changed dramatically beginning in the late 1950s, thanks to an initiative by the New Hampshire Fish and Game Department. Using sportsmen's dollars as well as matching federal grants, the department began constructing fishways on the coastal rivers using its own construction crew.

The first fishway was constructed on the Winnicut River in Greenland in 1957. The real boom in fishway construction got underway in the late 1960s. The Pickpocket Dam fishway built in Exeter on the Exeter River cost fifty thousand dollars, typical of the costs at that time. In less than a decade fishways were constructed by the Fish and Game construction crew on the Cocheco River in Dover, Exeter River in Exeter at two locations, the Oyster River in Durham, the Lamprey River in Newmarket, and the Taylor River in Hampton.

Coupled with the federal Clean Water Act of the early 1970s, which required towns to clean up the rivers, the fishways have caused a phenomenal turnaround in herring numbers.

A testament to the success of these fishways is the spawning run of herring in 2003. In total 194,116 herring were counted ascending the coastal rivers! Yes, counted by the Marine Division staff of the Fish and Game Department. They have been counted since the 1970s. Some fishways are equipped with special electronic counters to tally the fish numbers as they make their way up the current to the resting pool at the top briefly, before exiting the fishway. At other sites herring are counted by netting them and releasing them above the holding pen, a job I have gladly been occasionally asked to assist. In 2003 the rivers had the following herring counts: Cocheco, 71,199; Exeter, 71; Lamprey, 64,486; Oyster, 51,536; Taylor, 1,397; and Winnicut, 5,497.

And it's not just river herring that use the fishways. Numbers of sea lampreys, shad, and even two Atlantic salmon ascended the fishways in

2003. Trout, bass, and other freshwater fish are regularly seen in them as well. American eels slither down them!

Of course, maintaining and adjusting water flows is critical to the success of these fishways. Most years, depending on river flows, the fishways are opened in the first week of April. Marine staff checks them nearly every day in order to adjust flows and remove debris, including live beavers and huge snapping turtles on occasion!

The success of the fishways has restored the ecosystem of not just these coastal rivers, but all of Great Bay and beyond. Within the last decade, ten pairs of ospreys and a fish-eating hawk have started to nest around Great Bay. They were gone, as well, for nearly one hundred years and only returned because herring, which is significant forage for them, had been restored. Striped bass and other game fish swoop into Great Bay by May to feast on the herring as well.

The returning adult herring are only part of the success story. Each of the nearly one million female herring may deposit as many as two hundred thousand eggs in the freshwater stretches of the rivers to hatch and spend the summer growing to three- to four-inch juveniles. Typically, only about one percent, or maybe two thousand of the eggs survive the summer before the young then migrate out to sea by early fall. But even at these small survival rates that means as many as two hundred million, yes, two hundred million juvenile herring swarm out to sea each fall to provide forage to all manner of sea life.

Still, much is left to be done. On all of the rivers except the Exeter, only the last dam before the sea has been equipped with a fishway. Most rivers have other dams upstream still blocking the potential of these rivers. Just two years ago the Salmon Falls River was equipped with its first fishway after decades of haggling over its construction with the dam owner.

*Over twenty years ago, the lower dam on the Winnicut was removed. First published in 2004.*

# Fishermen Catch Tons of Tasty Treats on New Hampshire's Tidal Waters

Few could argue that New Hampshire's smelt is some of the best wild table fare available anywhere. Fresh fried coastal smelt are simply delicious and have been a favorite winter table fare on New Hampshire's coast for generations. And this state's fishermen literally catch tons of them in any given winter season. In 1981, a record twenty tons (41,394 pounds to be exact) was estimated to have been caught over the winter. The hotspot for the last several years has been the Squamscott River in Exeter. In 2004 an estimated six tons of fish were caught in the Squamscott!

Since 1978, the marine staff of the New Hampshire Fish and Game Department has been conducting annual winter smelt surveys. Marine staff members have checked preferred smelt fishing locations on Great Bay and its tributaries several times a week, except for a four-year period from 1983 to 1986, and in 2002 because there was no ice that winter. The fishermen surveys have provided all sorts of information. The survey takers query all the fishermen they find at each spot, gleaning such information as the number of fishermen, the number of fish caught, and the amount of time each fisherman has been on the ice. A sample of fish lengths are gathered as well as a scale sample to age numerous smelt.

As can be expected, the amount of fishing pressure and numbers of fish caught vary quite a bit from year to year over time. Ice condition varies tremendously on Great Bay from year to year. For instance, over the last five years the number of fishermen checked has varied from 1,123 in 2003, a year when safe ice came early and seemed to stay forever, to only five anglers in 2002, one of the mildest winters on record. And it's not just the ice conditions that cause major swings in the success of the fishermen; the fish population too is subject to significant changes over time. For instance, there was a die-off of eel grass from a disease in the late 1970s that probably impacted the smelt population, as likely did a major oil spill about that same time. Over time the very bottom of Great Bay may change too with new channels forming and old ones filling in. It may take a year or two for fishermen to locate the next hotspot.

Over the last quarter century of the study the smelt population has cycled at least four times up and down. In fact, a declining population in the late 1970s brought about more restrictive fishing measures that are still in place today. Basically, the netting season was restricted with most of the fishing pressure forced to the winter ice fishing only. A management plan written by the Marine staff in 1981 set three goals: to maintain or increase sea-run smelt; to provide for a recreational smelt fishery; and to provide for a commercial smelt fishery.

The success of this plan is really evident in the tremendous use of the sea run smelt by New Hampshire's fishermen. Figures from the last three or four years show smelt fishing continues to be alive and well. For instance, last winter an estimated 163,091 smelt were caught. The surveys say 75 percent of them were taken on the Squamscott River, 5 percent on Great Bay, 5 percent on the Lamprey River, and 15 percent on the Oyster/Bellamy Rivers. That's about eight tons of sweet success by fishermen. And there's tons more information too, such as it took 16,903 trips and 32,389 hours to catch that 163,091 fish at a rate on average of five fish per hour. Scale samples showed 67 percent were two-year-olds, 29 percent were three, and 4 percent were four.

Gee, it seems like the only ones with a little more patience than the fishermen sitting on the coastal elevator ice are the marine biologists who need to make sense of this data. It looks like the marine biologists plan to keep the fishermen doing what they like best: "keeping on, keeping on fishing!"

*In the last decade, due to lack of ice, there has been very little ice fishing for smelt or any other catch.*

*First published in 2003.*

# What's Up with the Clams Down on New Hampshire's Seacoast?

"Where have all the clams gone?" is a frequently asked question by red-faced clam diggers that have braved the winter cold and traversed the icy cold waters of Hampton Harbor. The answer is quite simple: "You should have been here yesterday," as the old saying goes. The fact is, it is the diggers themselves who have removed thousands and thousands of buckets of clams since the flats were opened, beginning with the Common Island flats in 1996 and the Middle Grounds in 1998. Both of these areas were closed to digging in 1988 due to problems with pollution. Once the closed flats were opened, clam digging was fantastic, the best it had ever been, and the word spread, bringing hordes of diggers to the flats during the annual nine-month season from September through May.

Since the 1960s Marine Division staff from the Fish and Game Department has monitored the Hampton and Seabrook clam flats. These clam flats show a textbook case of a predator/prey (man/clams) relationship with an upward cycle when there were too few clams to make digging worthwhile, so license sales plummeted following a collapse of the clam population. As digger numbers dwindled the clams were able to make a comeback, which again brought out the diggers to once again deplete their numbers.

This cycle has been evident several times in the last thirty years. The 1988 closure allowed a significant rise in clam numbers that now have been dug up and carted away by licensed diggers. Between 1998 and 1999 clam license sales went up by nearly a thousand, from 2,355 to 3,217, once the word of the great clam digging got out to the public. Conservation officers regularly patrol the clam flats during the weekly open days on Friday and Saturday. Even with the depleted number of clams evident this fall, hundreds of diggers show up to dig. On November 24, 585 diggers were counted. Between four hundred and six hundred diggers regularly show for a tide.

"Without a doubt most adult clam (over two inches) mortality comes from the removal of clams by diggers or from seagulls picking them out of the back dirt," says marine biologist Bruce Smith of the Fish and Game Region 3 Office in Durham. Juvenile clams, or clam spat (eggs), settle out of the water each year to begin a new crop of clams. Significant numbers of the juvenile clams are eaten by crabs and clam worms but once they are adult size, over two inches, there is little natural mortality. A disease of clams called neoplasia does occur but has not caused significant mortality as of yet.

Generally, it takes four years for clams to reach the two-inch size, which is the smallest most diggers will keep. The majority of the clams taken the last several years, since 1998, were larger-sized clams and were probably seven to ten years old. Although a significant number of the adult clams have been removed from the flats, this should not affect reproduction as the majority of spat probably comes from out of the harbor sources. In fact, clams are everywhere, including in the deep water never exposed at low tide. There is a reservoir of adult clams available for breeding and always will be.

Chief of the Marine Division John Nelson said, "It is the two-legged predator (man) that has the most impact on our clams. All phases of the health of clams are monitored and have been for a long time." He also noted that the Fish and Game Department has been working with other state and federal agencies, as well as the towns, for the long term to clean up the pollution affecting the flats. "This has been a priority," he said. He hopes other areas can be cleaned up to spread the effect of clam digging out to a much bigger area so the impacts on the current areas would be reduced. Although John did say, "We recognize the cycles and will take action if needed."

Recent surveys of the clam flats do show fewer numbers of clams compared to 1998, but they remain relatively abundant on an historical basis. For instance, during June 2000 there was determined to be about seven clams per square meter in the flats at both the Middle Ground and Common Island. This count is well above the historical lows in 1975 to 1979 and 1985 to 1988. In fact, in 2002, the current clam density is considered "robust" compared to long-term numbers. So even though the clam digging is not as good as when it first opened in 1996, it still is relatively good.

*Due to harvest cycles, over the last few years (the 2020s), the clam numbers seem to be dwindling.*

*First published in January 2001.*

# Global Warming Threatens New Hampshire Hunting and Fishing

During the last half century New Hampshire has witnessed a remarkable restoration of its fish and wildlife resources. For over the last three decades I was fortunate to help play a part of this restoration effort as a wildlife biologist for the New Hampshire Fish and Game Department until I retired in June 2007.

We have fish and wildlife populations unimaginable a century ago. Beavers had been trapped out by the mid-1800s and along with them most other furbearers. Moose were essentially gone as were turkeys and perhaps twenty-five thousand deer remained, and bears were believed to be fewer than five hundred. Dams and pollution wiped out fish populations in our rivers—especially shad, river herring, and salmon—that needed to reproduce in freshwater after maturing at sea.

There has been a dramatic reversal in fish and wildlife populations all across New Hampshire in the last fifty years, particularly the last three decades. For instance, when I started at the Fish and Game Department in 1976 there were fewer than one hundred turkeys in the state. We now have over thirty-five thousand. Moose numbered a few hundred and have grown to nearly six thousand. The bear population was pegged at barely over a thousand and has grown to five thousand. And even deer numbers doubled from a population of about forty-four thousand to ninety-eight thousand by 2007. The Federal Clean Water Act of the 1970s helped clean up all of our rivers that were once merely open sewers. The cleanup coupled with the construction of fish passage facilities on our coastal rivers such as the Cocheco River in Dover and our two major watersheds, the Merrimack and Connecticut Rivers, has brought about the return of shad, herring, and even some Atlantic salmon.

Global warming puts much of this restoration effort at risk. Quite simply, a warming earth stresses our fish and wildlife populations in many ways. For example, global warming could threaten our deer herd. Well, you might ask, "How come deer can live in the south where it is always warm?" Deer certainly live where there is no snow or harsh winter conditions. And at first blush you might think global warming will be good for our deer. Think again. When we have winter conditions of periods of deep snow or near zero temperatures, most of this state's deer seek shelter under hemlock trees. Sure, maybe winters will be shorter under global warming, but are predicted to have more extremes as well. There are 210,000 acres of hemlocks in New Hampshire, mostly in central and southern New Hampshire where deer densities are by far the highest. Just so happens just to the south of us there is a nasty little pest called the hemlock woolly adelgid that is killing off whole forests of hemlock. It is our *current* cold winter conditions that seem to be keeping this pest at bay. Warm up our temperature and wipe out our hemlock forests and the deer herd to boot. Moose are constantly threatened by the winter tick. They too seem to be kept in check by a harsh winter. And just a warmer temperature will stress our moose. Let alone more disease.

Our coastal river herring populations already seem to be threatened by the several last years of record warmth in the state. Warmer summer waters in some of our coastal rivers like the Taylor in Hampton Falls and the Exeter River in Exeter have caused a dramatic decline in spawning river herring. Between 1994 and 2000, ten thousand to forty thousand river herring returned each spring to spawn. That number dropped to only seven thousand in 2001 and just 147 herring in 2006. Unusually warmer summer river conditions are *already* affecting fish populations in New Hampshire.

As a wildlife biologist and more importantly as a New Hampshire hunter and fisherman I am deeply concerned with global warming's impact on my hunting and fishing. After all, we know how the stripers like those herring.

And I am not alone. A survey of New Hampshire hunters and fishermen by the National Wildlife Federation in September 2007 shows I have great company in my concerns about global warming. For instance, the survey found that almost three out of four (73 percent) hunters and fishermen agree that global warming is currently occurring. Two-thirds of the sportsmen believe humans are causing global warming. The survey also found that a majority (57 percent) believe the United States is doing too little to address the issue. In fact, 65 percent of the fishermen and 53 percent of the hunters believe global warming is a threat to their sport. While we see global warming as a future threat to our sports, a third of the hunters and fishermen say that global warming has *already* changed wildlife or habitat in their area.

Unfortunately, there is no one easy answer to the threat of global warming. Much can be done by each of us as conservationists and clearly much more needs to be done, sooner rather than later, on the national front.

*As of 2024, the state's turkey population has grown to over forty thousand, bear to over six thousand, and deer to over one hundred thousand. New Hampshire's moose numbers have dropped in half to about three thousand, driven by climate change.*

*First published in October 2007.*

# If Only Moose Could Vote

If only a moose could vote. If I was a moose, I would vote for clean air and the new Carbon Rule that is part of the Clean Air Act. You see, New Hampshire moose are in trouble, real trouble. In fact, we have been for some time now. Air pollution is delivering a one-two punch to us. This all started back some twenty years ago when humans first realized that their air pollution, including acid rain and air pollution like mercury and cadmium, was accumulating in our bodies. You know, the stuff you shoot up into the atmosphere from coal-fired power plants and similar dirty factories. It didn't take you long to find out that me and my cousin, whitetail deer, had accumulated this toxic brew in our livers and other organs. Yes, you have poisoned us enough to make us too toxic for you to eat part of us. You humans have depended on moose and deer for food and clothing since the beginning of time. And this is how you repay us?

Now this same pollution is causing the earth's temperature to rise. You call it climate change or global warming. We call it the kiss of death. To put this all into perspective let's look back a few decades. We moose were practically gone by 1901 when we first received protection from unregulated hunting. And from fewer than fifty moose by the mid-1900s we thrived and grew in numbers. By 1988, the first hunting season was established with only seventy-five hunting permits allocated. And still we thrived and grew in numbers. Moose hunting permit numbers followed us up as our numbers grew. By 1991, one hundred permits were issued. Our numbers grew even more, and by 1997, 570 permits were then issued. But the earth was warming, increasing on average one degree each decade since 1970. Yet our numbers continued to grow as did the moose permit numbers. Yet things were changing as our winters warmed. Our nemeses, the moose ticks, were growing in numbers with ever-warmer winters. By 2005, moose numbers grew to some seven thousand in New Hampshire. And hunting permits swelled to 675 by 2006. But then came even warmer winters that really caused moose tick numbers to thrive.

Most years a moose like me can live with upwards of thirty thousand of these winter ticks biting, itching, and burrowing into my flesh as they suck my blood. But given the extremely warm winters of late, due to all that air pollution, tick numbers have swelled to sometimes 160,000 ticks on a moose like me. This is a number that kills us slowly, as we scratch and rub ourselves against trees to try to scrape the tens of thousands of ticks off, along with our protective fur. Then we slowly suffer and die of hypothermia.

In 2007, records show another unusually warm winter, and moose numbers fell to an estimated population of only 5,500. And with that downturn in our numbers the moose hunting permit numbers were reduced to only 515 for 2008 and 2009 hunting seasons. This was followed by more declines in moose numbers and permits the following two years of only 395 permits.

New Hampshire Fish and Game moose biologist Kristine Rines figures that upwards of 40 percent of moose mortality is from winter ticks on average. But after a winter like 2010, which was extremely mild, Kristine estimated over 20 percent of us adult moose and all our calves died the following winter, 2011, from winter ticks.

Now going into the 2012 season, Kristine estimates that our moose population is closer to four thousand. And hunters will see a much further reduction in moose permits to only 275 permits to be issued for 2012 and 2013.

So thanks to a warming climate, especially too-warm winters resulting in a scourge of winter ticks, our moose numbers have fallen from seven thousand moose to fewer than four thousand today, a 40 percent decline. Moose hunter permits will have fallen 60 percent from 675 to 275 for the 2012 season. A recent quote in the *Northern Woodlands* magazine by Kristine Rines said, "Moose are facing a triple threat in our changing climate. Increasing temperatures, changing forest species and increased mortality due to parasites may make it very hard to maintain a viable moose population in New Hampshire in the future."

And it is not just in New Hampshire that moose populations are being affected by too-warm winters. In northern Minnesota moose numbers are down some 90 percent as the winters there have warmed about four degrees on average over the last three decades. Maine moose too seem to be taking a hit from high winter tick numbers. Maine shed hunters, hunting for antlers, reported discovering hundreds of dead moose there last spring as well.

Could New Hampshire's moose be essentially gone within another decade? If only moose could vote. If I was a moose, I would vote to support the Environmental Protection Agency's efforts to curb air pollution like mercury, cadmium, arsenic, and carbon. I would vote for our senators and congressmen who support the new Carbon Rule part of the Clean Air Act. I would call them to ask for their support. Until you see my stubby tail sticking out of the voting booth next to you, can I count on you? Can I count on your votes and calls to support the Clean Air Act? I'm counting on you.

*Moose have continued to decline and there are three thousand remaining in New Hampshire in 2024. Our efforts to lobby for the Carbon Rule were not successful; however, the Clean Air Act is still active.*

*First published in 2012.*

# Downwind and Dirty: Living in the Shadow of a Coal-fired Power Plant Stack

I have lived in the shadow of the Bow, New Hampshire, Merrimack Station coal-fired power plant smokestack for thirty-five years. I first lived as a resident of Allenstown, where I moved in 1974, then in 1979 to the house I now live in overlooking the Suncook River in Epsom. The top of the smokestack at the power plant is visible not far from my house if I look across a nearby snow-covered cornfield. At a glance I can gauge the likely power output by the plume of smoke shooting my way from the snout of this ancient dragon. Yes, the Merrimack Station is over forty years old as is the technology on which it was built. Technology from the middle of the last century when clean water was of little concern, let alone clean air.

Now, in 2009, New Hampshire stands at a crossroad on how our future energy needs should be met. There are plans to develop a huge wind energy plant in the northern region of the state. At the same time Public Service of New Hampshire (PSNH) is moving forward with plans to spend some say upwards of over a billion dollars to keep the Merrimack Station in operation another twenty years or more; all the while it will continue to spew tons of pollution into our air each day. We have faced similar challenges before when our rivers were no more than open sewers.

I grew up in the sixties in Londonderry just south of Manchester and I bore witness firsthand to how little concern there was about water pollution. I would fish the local Watts Brook for native and stocked brook trout but was always amazed at the filth of the Merrimack River, where the brook's clear waters emptied into its chocolate-colored bowels. Yes indeed, the Merrimack was nothing more than an open sewer with confetti-colored toilet paper flowing by with all the human waste churning in the river's current. It took the federal government to pass the Clean Water Act in the early 1970s to clean up the Merrimack River as well as the Suncook River I now live on. Folks here in town talk about the old days when the Suncook River ran different colors depending on what color dyes were in use at the tannery in Pittsfield. Decisions were made forty years ago to commit the federal, state, and local governments to clean up our rivers.

While we have made great strides in cleaning up our rivers and streams, we seem to be treating the very air we breathe like the rivers of the past. Did you know that the Merrimack Station is the single largest point of global warming gases in New Hampshire spewing out 3.7 *million tons* of $CO_2$, about 20 percent of the state's total, and 120 pounds of mercury *per year*.

Over my thirty-plus years as a wildlife biologist in New Hampshire I have come to know just how these numbers are impacting our fish and wildlife.

For instance, I happened to have a friend working at the Department of Environmental Services (DES). He was to begin an assessment of pollution in fish in northern New Hampshire but wanted to practice the testing procedure first and asked me, as a fisherman, if I could get him a couple fish to practice on. He needed predatory fish like pickerel or bass, which are at the top of the food chain. So I gave him a foot- to fourteen-inch-long pickerel from Little Durgin Pond in Northwood and a similar-sized largemouth bass from Beaver Pond in Bear Brook State Park. He called me a few weeks later in a rather excited voice: "Wow, those fish were off the charts!" he exclaimed. "These fish have the most mercury in them of any other fish ever tested in North America. Just one meal of one of these fish by a pregnant woman could potentially affect her fetus."

And it is just not fish that I have been involved with testing. In another example, as a biologist for the Fish and Game Department I was involved with submitting pieces of moose livers for testing. Here too the pollution from our power plants has caused acid rain that has stripped heavy metals from the soil, causing moose to take up high levels of cadmium in their livers. Too high for human consumption these days, it seems; same goes for deer livers too.

In a very recent discovery, it appears that acid rain is stripping aluminum from the soil as well as sending it into the tributaries of the Merrimack

River where we have a decades-long effort to restore Atlantic salmon. Studies suggest that aluminum is coating the gills of the juvenile salmon. This seems to be no problem in their two-year stay in fresh water. But as soon as they molt and migrate out to sea, they suffocate in salt water because of the aluminum. Each year one and a half million salmon fry are stocked into the Merrimack River system as part of a four-decade restoration effort. While the blame in the past has been placed on the unknown events at sea, the problem now seems to be right here in our own backyards. The plume of effluents from our power plants casts a huge shadow over much of the fish and wildlife that we are just now learning about.

I also collected mink carcasses from New Hampshire trappers and sent them off to the US Fish and Wildlife Services a few years ago as part of a study to look at mercury in both mink and otter. Here again mercury levels in the mink I collected were very high. According to my friend at DES, "The levels of mercury in the mink you submitted would be harmful to a human."

In a recent proposal released by PSNH they are planning to spend a half billion dollars in adding a scrubber to the Merrimack Station to reduce mercury levels by 80 percent. Just two years ago this same scrubber was to cost only two hundred and fifty million dollars. In another outside estimate the cost of this scrubber will more likely be $1.2 billion. Just this February of 2009, the EPA has announced that they will issue at least a 10 percent tighter mercury control requirement for coal plants that will be effective in 2010. The current proposed scrubber does *not* meet the new proposal. Besides, how much mercury is "safe?"

New Hampshire senator Harold Janeway has introduced a bill to require the state's public utilities commission to determine just what the costs will be to keep the Merrimack Station operating for another twenty years and whether there are better alternatives to spending potentially billions of dollars on a plant that in the best of condition will continue to spew vast quantities of pollution into New Hampshire's air, the air you and your grandchildren must breathe.

*In 2024, plans were announced to close the Merrimack Station coal plant by 2028.*

*First published in February 2009.*

# Clean Water and Air is a Must for Our Fish and Wildlife

It seems like we all take for granted our clean rivers and clean air. It wasn't always that way. It was not that long ago things were very different. Yes, when I was a teenager growing up in Londonderry in the sixties, things were very different with our rivers. My best friend Rick and I would fish Watts Brook for beautiful brook trout, wading and splashing in the cool brook as we made our way downstream all the way to where it dumped into the Merrimack River in Litchfield. Here was a scene of a far different world. You couldn't stand on the rocks in the river as they were covered in sewage. Yes, human waste filled the river as far as the eye could see. Pure raw sewage.

And the Merrimack was filled with sewage from not only Manchester but Concord too, and beyond. I remember going duck hunting on the Merrimack River on the north side of Concord with a college (UNH) friend in 1970 or so. Here too the river was full of human waste. Folks who I know here in Epsom remember a time the Suncook River that flows past my house used to turn different colors depening on what color die was used at the tanning plant upriver in Pittsfield. I remember all our rivers in this part of the state were pretty much to be avoided as far as fishing or swimming goes.

Thankfully, all this was to change thanks to the Clean Water Act of 1972. This act gave the Environmental Protection Agency (EPA) the authority to begin the task of cleaning up this nation's waterways. As my good luck would have it, I actually paid for my college education by working construction, summers. Some of that work included installing sewer separation pipes in Manchester and later on sewer pipe construction in Hooksett. So the Clean Water Act actually helped me pay for college. Another clear advantage to cleaning up our rivers was that it provided good jobs.

The Clean Water Act worked just great at getting our rivers cleaned up and helped protect against further pollution. Except the definition of what "waters" meant came under dispute over the following decades. Originally the "waters" were described as "All waters with a significant nexus to navigable waters." For over three decades most streams and rivers were considered protected under the act.

However, Supreme Court rulings in 2001 and 2006 muddied the waters, so to speak. Well, no, the new court definition actually did muddy some waters. The new court rulings took away the protection of the smaller tributaries. Mostly I think of them as our native Eastern brook trout waters. These are the flows that trickle through our forests much of the year, but in a hot dry spell much of the stream bed nearly dries up, leaving just the deeper holes carrying our native trout through the drought. These are the small streams that as a boy, I could leap across, and had ankle-biting cold water but held the most beautiful of trout. In fact, 60 percent of America's streams are seasonal or temporary. The 2006 Supreme Court ruling defined "waters" as "relatively permanent, standing or continuously flowing bodies of water forming geographic features as streams or oceans." This would be the kiss of death for many of our local trout brooks.

Fortunately, the EPA has moved to restore the Clean Water Act's protection of our smaller brooks. The new rules define "waters" as "any tributary showing significant features of flowing waters."

This definition will clearly protect what most of us know are our native trout streams. But we must ask our congresswomen to support this definition and not overturn the EPA's effort to protect our smaller tributaries.

Secondly, we know that moose numbers are down significantly in New Hampshire due to climate change. Thankfully, the EPA's Clean Power Plan is set to curb the release of carbon over the next three decades by some 32 percent. Just like the Clean Water Act, this too will bring jobs in the clean renewable energy and energy conservation fields. Just recently Fish and Game biologist Kris Rines said, "If we don't wrap our arms around what we need to do to reduce climate impacts, we need to recognize that our entire world will change." Kris has been tormented as she has monitored a nearly 50 percent reduction in moose numbers over the last decade and a half as our New Hampshire winters have shortened. As she has frequently said, "We need snow on the ground in April when the female ticks fall off the moose if we are going to maintain our moose population." The trouble is our winters have warmed some four degrees on average over the last forty years, according to a recent UNH study.

Here again it is up to us to make sure we are contacting our senators with a request for them to support the Clean Power Plan.

*The warming trend continues as of 2024; the winter and summer in 2023–2024 had record-breaking heat.*

*First published in 2008.*

# Do Bears Hoot?
# Bear Myths and Medicine

Bears have been animals of myths, mystery, mystique, and medicine for millennia. All across the northern hemisphere, native peoples revered bears as the holders of great powers. Indeed, bears were believed to die each fall (when they denned) only to be reborn each spring. They were the "keepers" of the earth.

Scandinavian legends speak of the ability of some people to assume the characteristics of a bear. Our English word "berserk" comes from this legend. It was thought that if a warrior donned a bearskin (called a bear-sark), he would be given the power, strength, and stamina of a bear. Native Americans had similar legends.

Did you get lost? Have you got your "bearings" straight? This is a word also related to the bears. The North Star, the tip of the Little Dipper's handle, and the tip of the tail of the Little Bear has guided travelers for centuries. Interestingly, the Big Bear, in which the Big Dipper appears, points the way to the North Star.

Bear parts and bear grease have long been used for medicinal powers. The early American settlers regularly applied "bear grease," the rendered fat from bears, to all sorts of ailments, such as arthritis. Some still believe in its use today.

The bile from bear gall bladders has been used as a "stomach medicine" for hundreds of years. In the early 1980s, two doctors from South Korea were doing research on the bear gall bladder as a medicine while at a hospital in Boston. The call went out for bear gall bladders. Well, frozen in the Fish and Game freezer in Concord was a whole cluster of road-killed bears. We used to sell them at the department's annual auction of road kills. After the bears were thawed a little, the doctors came up and removed the gall bladders. Their research found that indeed, bile from a bear, especially one near hibernation, could effectively dissolve gallstones. The acid, named after bears, is now synthesized and used to prevent gallstones.

Indeed, bears are good medicine for the body and the soul.

By the way, there is a long-held belief in New England that bears "hoot" in the spring to communicate with each other, especially during breeding season. This is a legend only common to the Northeast and does not seem to be supported by scientists.

*First published in 1991.*

# Go Take a Hike . . . at Night

I just love to be in the woods day, or night. Yes, I have hiked and worked in the woods at night much of my life.

I think nighttime is the best time to really get to know the woods and your ability to sense things in the woods. Let's face it, 99 percent of what we sense during a daytime hike is with our eyes. We really limit the use of our senses with a day hike. I liken it to the difference between learning a pond by taking a canoe paddle across the water or diving in and swimming across to learn how it feels.

Night hiking frees up all your senses and gets your blood flowing to them to feel, hear and smell your surroundings.

It is amazing how acute your sense of feel becomes in the dark. Now you can sense each step as you walk along. All good practice for those fall days when you want to sneak up on the bedded buck. And the sense of feel blossoms across your body. You can actually feel a slight breeze on your face or other exposed skin. Here again training yourself to keep track of the wind while deer hunting by day.

And do your ears perk up to the slightest change in sound? It's like you can sense the woods around, how close to the vegetation you are hearing it as you approach it. Let alone distant sounds that are amplified in the dew-filled night air. Sounds abound at night.

And let's not forget your sense of smell, now ever more acute in the dark. Yes, your nose actually can be used to your advantage. I have picked up the smell of moose, deer, bear, and fisher before in the woods once I "learned" what they smelled like. Yes, I have had the advantage of being a wildlife biologist and have learned these fragrances while handling live animals, but any deer or moose check station offers ample opportunities to sample your own. Fisher have a distinctive sweet musky smell that is easy to learn. You might check with a trapper to get a whiff of one.

Now to take a hike at night I don't suggest you blunder into the wilds without a flashlight. By all means have one along. And a compass and map should be handy . . . just in case. Yes, night woods can be very confusing. I suggest you start out in woods you are very familiar with already and stick to the main woods roads and trails. Why not take that GPS along and don't forget to click in your starting point. I find not using them and just relying on your good wood sense will keep you safe and headed in the right direction. After all, using them requires light and your eyes prefer a day hike. But you are on a night hike.

I have found myself in some of the wildest parts of New Hampshire at night while working as a wildlife biologist for Fish and Game. For instance, I worked on Long Island on Lake Winnipesaukee on a deer removal project in the eighties. Biologist Henry Laramie and I removed twenty-five deer over a two-week timeframe one fall. We tried several trapping attempts with some success, but the most efficient was when we used our tranquilizer guns with a radio-fitted dart. Yes, I got to hunt deer at night. Generally, we were driven around the island by a conservation officer until we got close enough to hit one with a dart. Work commenced at dusk and we worked into the night, sometimes all through the night. The darts at that time were made of aluminum and flew like a rock. So you had to be pretty close and kind of lob them into the deer's butt. And that was the easy part. My job was to find the hopefully sleeping deer fifteen minutes later when they were down. But any light or sound might trigger them to get up and run off. So I listened to the beep of the dart with headphones as I slowly and quietly stalked the downed deer in the dark.

Once, I had just put a rope around a deer's neck, but before I could tie it off, it ran off trailing the rope like a noose. I waited again, and unfortunately had already pulled the radio dart out, so I had to crawl in the dark on my hands and knees until I felt the quarter-inch rope on the forest floor.

I had this same duty while radio collaring moose in 1986 and 1987 in Pittsburg. I was the "moose finder" once we had a dart in one. I spent hours in the dark in Pittsburg stalking downed moose, over forty of them in two falls mostly without shining a light. Yes, I am very comfortable in the dark without a light pretty much anywhere. And you can be too. So get out at night for a hike. Maybe let someone know where you will be and when you plan to return, just in case. But nothing trains all your senses like a night hike. I bet you'll be a better deer hunter for it.

*First published in 1998.*

# New Hampshire Stripers Are for the Birds!

New Hampshire stripers are for the birds! A cloudy but warm wonderful day on Great Bay recently sure proved it to me. You see, my lifelong friend and frequent fishing partner, Rick Hamlett, and I spent the day chasing birds around Great Bay all the while hooking stripers. In fact, we brought at least a hundred stripers to my boat during the five-hour marathon bird and striper chase.

Serious striper fishermen know that the very first thing you need to do when on the water is watch for birds, especially diving and feeding birds. From their vantage point in the sky, coastal birds simply cruise until they see a school of fish below, then dive into them grabbing what they can. It typically is a feeding frenzy for the birds and simultaneously for the stripers chasing the school of bait fish. There is simply no better way to find feeding stripers than watching the birds.

And it's just not the diving feeding birds you need to watch. I've noticed the birds sometimes simply fly over the schools of bait, often acting nervously but not diving. Perhaps at these times the bait fish are simply too deep for them to catch. All I know is the birds always know where the bait is. Find the birds and you'll find fish!

In New Hampshire the birds to watch for on the bay and elsewhere on the coast are common terns. Actually, the common tern was *not* very common until just recently. In fact, they are a species of concern by the New Hampshire Fish and Game Department.

Like so many things in a small state like New Hampshire, everything seems to be connected one way or another. My connection to the terns in the state began over a decade ago. As a wildlife biologist for Fish and Game for nearly thirty years I have been involved with all sorts of projects. In the late 1980s, the New Hampshire Audubon Society biologists were trying to count nesting common terns. Since 1988, I have been the regional

*Eric Orff with striped bass catch.*

wildlife biologist out of the coastal office in Durham, which also hosts staff of the marine division. Annually, I would commandeer one of the marine division's boats to assist the Audubon biologist in searching our coast for nesting terns.

Most years we found anywhere from five to eight nests on a couple of the islands in Great Bay and for a few years we had as many as twenty-five or thirty nesting pairs in the Hampton Marshes. Decades ago, the terns had been killed off or displaced by the burgeoning gull population along the coast. The gulls prey on the tern eggs and young despite a fierce fight by the tern parents. When tern colonies have dozens or hundreds of pairs, they can usually keep the gulls at bay. But when they are few in numbers, they fall victim to the gulls. Our open dumps for decades have helped swell the gull population and caused them to drive the terns off the coastal islands, especially out at the Isles of Shoals.

All that changed in 1997 when the Fish and Game Department teamed up with New Hampshire Audubon for the first-ever attempt to restore common terns to White Island out at the Isles of Shoals. Here too, I played a small part and did for several years. I was called upon to transport crews and equipment out to the island using the marine craft again. Now let me explain to you that basically I'm a chicken when it comes to taking any craft that far out to sea. Sure, I've been around boats all my life, but for the most part I'm a sweetwater man. And of course I had to deliver the crews and equipment in a not-so-big Fish and Game boat in April.

It was in April 1997 that I first conveyed crews to stay on the island to drive the gulls away, while blaring tern calls 24/7 through loudspeakers and shifting the dozen wooden "decoy terns" around frequently to make them appear lifelike. But it worked! That first year a half-dozen pairs successfully nested thanks to the babysitting of the staff who lived out there all summer.

In just six short years the common tern colony has grown from the half-dozen pairs to 2,414 in 2003. Last year they successfully fledged 3,212 chicks. More astonishing is the fact that sixty-three rare federally listed roseate terns nested, producing fifty-six chicks as well as the even more rare arctic terns, that produced six chicks from six nests. A late June report by Fish and Game staff living on the island indicates similar numbers of terns for 2004.

Now back to the stripers. Often fishermen and hunters wonder why Fish and Game folks even bother to work on such silly things as terns. Why? What good are they? If you are striper fishermen you won't be asking that question. For me, helping in just a small way to bring back the terns, has improved my striper fishing immensely. My nearly thirty years of experience as a wildlife biologist has shown me over and over again how fish and wildlife are intricately connected. As for striper fishing, I'll look for the birds to be my guide.

*There currently are over 3,000 pairs of common terns nesting at the Isles of Shoals.*

*First published in June 2004.*

# Guess Who's Coming to Dinner?

*Honk, Squawk, Yap, Squeak, Splash, Flutter, Flutter, Flutter . . .*

All eyes are peeled outside and the questions begin when I arrive home from work with a particular grin on my face. What did you bring home today? Is it a baby? Or injured? How many? The questions remain the same after five decades of my life. Only the faces have changed from my mother, father, brother, and sister to my wife, daughter, and son.

I have been intrigued and amazed by fish and wildlife all my life. I grew up relentlessly pursuing wild things and generally bringing them home. I knew when I was four years old and catching horned toads and tarantula during our brief stay in Oklahoma, before returning to Maine, that someday I would work with animals and become a biologist. Now, I am a biologist and have been one for nearly half my life. But old habits are hard to break. Wild animals still just seem to follow me home. Except the size and variety has increased in proportion to my size. Yup, I'm grown up now. Bear and deer have been guests in my home and who knows when the moose will arrive!

I have been a biologist for the New Hampshire Fish and Game Department since 1976 and since February 1983 I have operated a wildlife control business, part time, out of my home in Epsom, called Bat and Wildlife Control Specialist. I have often been in a position to evict someone else's unwanted wild guests from their home, then bring them to my home. Or have been in the office on a late afternoon when some orphaned or injured animal has arrived at the Fish and Game office. Our house welcomes all guests. Generally in a couple of days, I am able to line up one of the state's licensed wildlife rehabilitators who will give these animals good long-term care. No one is allowed to keep any wild animal permanently. Wildlife belongs in the wild. The goal always has been to get healthy animals back into the wild as soon as possible. Here are some of the animals who have been my guests.

Bears generally arrive in two different sizes. Newly born orphaned cubs arrive midwinter when a mother and young are disturbed in the den and the young have been abandoned. These are helpless little creatures about the size of a large kitten. Cubs require intensive care, bottle feeding every few hours. The house is abuzz when I bring cubs home. Yearling bears show up in January in years when the normal fall foods for bears such as nuts and berries are absent due to weather conditions. These bear should have gained enough weight in the fall to carry them through hibernation. The malnourished bear wake up early and wander, sometimes right to my doorstep (via some goodhearted citizen)! Starving yearlings just need a dry pen and food as they skip hibernation. I have been able to transfer these animals to a good caregiver within a few days. They always do well and are released in late spring, often right back where they were found.

For several years my house served as Grand Central Station for fisher. The New Hampshire Fish and Game Department had teamed up with this state's trappers several times to relocate live animals to other states for reintroduction efforts. We were collecting live fisher to send to Connecticut in the early nineties and between 1994 and 1997 we collected and shipped another 175 fisher to Pennsylvania. December evenings found me in my garage until late at night securing the fisher I had collected that day into special plastic transportation tubes. I left my house at 4 a.m. the next morning and delivered the fisher to the Manchester Airport for the early flight to Pennsylvania. I knew these fisher well! They made great guests. Despite their undeserved nasty reputation, I found wild fisher one of the easiest animals to handle while in captivity. They were very shy and appeared mild-mannered when in a cage.

This was not the case for the few river otter I held in my garage until they too were shipped to Pennsylvania about ten years ago. Otters fought the cage and acted aggressively whenever anyone approached their cage. The otter stayed for a number of days, unlike the fisher who were just overnight guests. Luckily when I built my garage, I put in a grease pit so I could easily work on my cars. Just like the pits in the zoos, my cleaned-up grease pit served as a holding facility for otters.

I had furry little baby foxes too. Their pale coats and grey eyes made them an awesome sight. One scampered under our dishwasher and it took me a while to get him out! A few animals I have had to keep all winter. An opossum had a lengthy stay in my basement, as did a poor sick Canada goose two winters ago. It took three days of force-feeding the goose to get her strong enough to eat on her own. How she honked when I let her go the next spring at Great Bay. The mink visited for a few weeks, as did the raccoons and ducks I've hosted.

A few times I have had to gather up several hibernating bats from houses and have brought them home. These stayed a while, a little too long for my wife's liking. While my intentions were good, my plan was flawed. Unbeknownst to my wife I simply put the bats in a paper bag, punched a few small holes in it, and hid it away in my cellar. After all, the bats were hibernating. Well, you know about the best-laid plans.... Somehow bats began leaking out of the bag during a warm spell; in fact they must have practically spilled out! Their favorite hiding place seemed to be the clothes washer. There was a while my wife was real reluctant to do laundry. She kept asking me how many got loose and I kept saying, "I think I got the last one." This went on all winter! That's okay; the next winter I hid the bats in a better paper bag in our bedroom closet. Way beyond her secret hiding place for the Christmas gifts. She never caught on and they didn't get loose. (Honey, there really are *no* bats under our bed this year.)

The pheasant episode should have been a short story too if it had not been for this particular pheasant. This bird was brought to Fish and Game headquarters in Concord and I volunteered to bring him home. He was emaciated and in terrible shape. Cracked corn put weight on him quickly, but this was a harsh winter so I decided to keep him until spring. By mid-April, the snow was gone and I opened his cage and watched him amble off. I never expected to see him again although I was hoping to hear him crow in gratitude a few mornings. This was not to be! Not only did he not leave but he moved right back into the yard and claimed it as his territory.

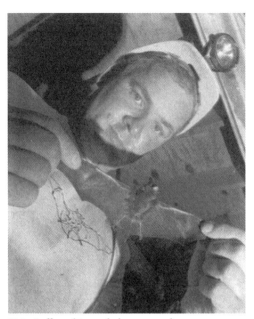
*Eric Orff working to help preserve bats.*

He became very belligerent—especially to my three-year-old son. One afternoon, we heard the kids screaming in the backyard and looked out to see them "treed" on the swing set. The pheasant was after them and the only thing keeping him at bay was the furiously swinging swing! I went right out and chased him off, or so I thought. For over a week he intimidated my two kids and at one point chased my son right into the house. I have pictures of this pheasant under my kitchen table! His undoing came a few days later when he challenged my neighbor's Buick.

Being in the wildlife profession has brought lots of opportunities to meet new friends and strange guests. Who knows what tomorrow may bring. By the way, I think I'll go out and visit the wild turkey who's spending this rainy late January night in a cage in my garage. I always provide room service.

*First published in 2001.*

# New Hampshire Wildlife Flourishing

For many species of New Hampshire wildlife, the current population levels and range are greater than in any other time in the last two centuries. Yes, centuries. Particularly the last several decades have seen a resurgence in animal numbers as modern wildlife management practices, including protection and regulated seasons, took place.

Many animal populations were nearly decimated by the late 1800s and the early part of this century. Land clearing, agricultural practices, and market and subsistence hunting and fishing brought species to all-time lows. Despite the near elimination of some species, protection by closed seasons did not occur for some until the 1930s. Animals considered pests or predators were unprotected or bountied. Bear were bountied until 1956 and bobcats until 1974.

The decade of the nineties shows most furbearer species at near all-time highs. It was estimated there were only 240 beaver in 1915. Beaver were relegated to as few as one or two colonies in northern Coos County along the Canadian border at the turn of this century. In 1905 beaver were given complete protection. As the number of beaver grew, some were captured and moved south, four were trapped in 1926 and moved near Plymouth, and in 1936 there were fifty-eight moved further south. By 1940, there was an estimated seven thousand beaver. Beaver populations built rapidly and spread all over the state by the 1960s.

Beginning in 1941, beaver trapping was allowed again, but with a very limited season and bag limit. As the population grew, seasons were lengthened to help control overabundant populations. By 1979, over sixty-five hundred beaver were trapped in a single year and valued at over a quarter million dollars to the trappers. Beaver are abundant statewide. The abundance of beaver and their capabilities of constructing dams and creating more wetlands has had paralleled benefits to many other water-dependent species. Population increases in otter, raccoon, waterfowl, herons, and moose can likely be directly tied to the beaver's success.

Fisher were nearly extinct by the time they were given protection in 1934. By the 1960s, they had once again spread statewide on their own. Seasons were again allowed and over eleven hundred fisher were taken in 1971. Fisher remain abundant statewide with four hundred to five hundred taken annually. Although two species of furbearers, the mountain lion and the wolf, were extirpated in the 1800s, two new species have since entered the scene, the opossum and the coyote. The opossum first appeared in New Hampshire with the capture of one on December 10, 1915, in Warner. This is a Southern species that has spread north into our state and has become very abundant in the southern half of the state. Although it occasionally occurs in Coos County, the cold weather and deep snow has slowed its expansion north.

The coyote took the opposite route into the state. The first coyote identified in the state was captured in Holderness in 1944. There were sporadic reports of coyotes in the state increasing slowly through the 1950s and 1960s. The march of the coyote really began about 1972 in the upper Connecticut Valley in Coos County. By 1980, a wave of coyote expansion had swept down the Connecticut River Valley and pushed eastward to the coast. By the end of the 1980s coyotes had probably colonized most of the suitable habitat in the state.

Some furbearer species have not fared as well. The pine marten has only gradually started to expand in central Coos County. Bobcat numbers have declined probably as a result of maturing forests and competition for prey from coyote.

Moose have shown the most phenomenal increase. A dwindling and sparse population was not protected until 1901. The moose was slow to return and by the late 1950s only twenty-five to thirty were estimated to occur. The seventies and eighties brought a phenomenal growth in the moose herd. This is reflected in the number of moose killed by cars each year. In 1981, fifteen moose were killed, doubling to thirty-one the next year, and doubling to sixty-two in 1986. By 1991, 197 were killed by cars. By 1988, a moose hunting season was established with very limited restrictions.

The moose herd continues to grow with moose reported statewide, and animals reproducing statewide as well. Moose are in our backyards and skirting the edges of our cities.

The whitetail deer population has gradually grown from a residual population in the 1920s. Deer have always been hunted, but the annual kill fluctuated between fifteen hundred and two thousand from 1922 to 1935. The deer kill peaked in 1951 at 11,462 and again in 1967 at 14,121. Much of this historic high kill came from the far northern part of the state. Severe winter weather, long hunting seasons, and changes in habitat caused a decline in numbers of deer. Enactment of limited doe harvests has brought the deer herd on a steady rebound. In 1992, over ten thousand deer were taken by hunters, the highest since 1968. A big switch has taken place in abundance levels of deer. Where once a large portion of the kill came from the northern big woods areas, the recent resurgence has been caused by record harvests in the southern counties. Record numbers of deer now occur in much of the southern and especially southeastern parts of the state. This trend is likely to continue. Record numbers of deer were taken by bow and muzzle loader hunters primarily from the southern third of the state.

Black bear now live in more areas of the state and in greater abundance than since the early 1800s. Bear were one of the last species to receive protection and management as a game species. Bountied until 1956 and not protected until 1963, bear were still not given big game status until 1983 and were turned over to the Fish and Game Department for management in 1985. Since receiving the authority to regulate the harvest in 1985, the department has closed the season in the southern half of the state to allow for a natural range buildup, especially in Sullivan, Cheshire, Merrimack, Belknap, and Strafford Counties. This buildup is gradually happening but will take many years to complete. The northern three counties of Coos, Carroll, and Grafton have always been the stronghold for bear and still contain over 90 percent of the population. Bear numbers were their lowest in the late 1800s when only thirty to a hundred were killed each year despite a bounty and no protection. Since protection began in the 1960s the bear population has doubled to nearly three thousand and provides an annual harvest by hunters of about two hundred and fifty bear.

Turkeys were extirpated over one hundred and fifty years ago and none were seen until the Fish and Game Department acquired wild turkeys from other states and transplanted them here in 1969 and 1975. Turkeys began to flourish after the 1975 transplant and offspring were relocated to seven other areas across the state between 1978 and 1986. Turkeys have flourished in many of these areas and allowed for a limited spring, male-only season in 1980 with thirty-one birds being taken by hunters. By 1985 the kill had doubled to 61, more than doubled by 1989 to 154, and doubled again to an all-time high of 352 in 1992.

These dramatic successes have also been mirrored by non-game and non-hunted species as well. Many species have flourished expanding their range into and within the state, thanks to changing habitat, protection of wetlands, and better environmental laws regarding pesticides and with the help of man.

Ravens, ospreys, cardinals, house finches, mockingbirds, mourning doves, and turkey vultures have either migrated into the state or have greatly expanded their breeding range within the state. The bald eagle has once again successfully nested for four years after a more than fifty-year absence. Osprey have begun nesting in the coastal region after a long absence. The peregrine falcon required the hand of man to reintroduce them into traditional nesting sites beginning in 1976. Since 1981 they have successfully nested, with a growth in nesting pairs producing eleven offspring in 1990.

Waterfowl populations have shown great changes. In the 1960s a relatively new species, the ring-necked duck, began nesting in the state. The wood duck, which was thought to be going extinct at the turn of the century, now flourishes and has become nearly the most abundant duck species in the state.

The last decade has brought a more urbanized population of mallards and Canada geese. Nesting populations of Canada geese have flourished in Hillsborough and Rockingham Counties over the last ten years. Over thirty-one hundred urban Canada geese were counted in the fall of 1992. Winter counts of resident mallards in January through March of 1993

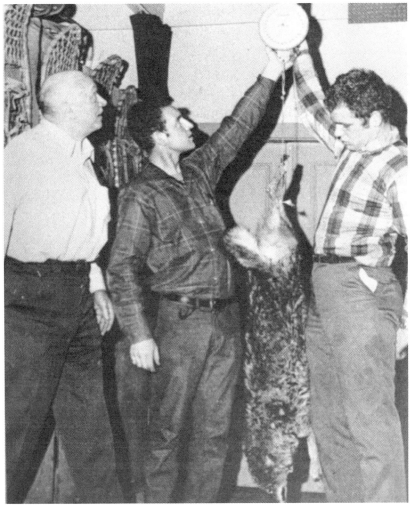

*Eric Orff helped conduct early research on Eastern coyotes while a student at the University of New Hampshire in 1969.*

showed nearly five thousand wintering from the lakes region south.

Flourishing wildlife has not just been birds and mammals, but fish too. Enforcement of new environmental laws such as the federal Clean Water Act coupled with the construction of fish ladders at century-old dams and the restocking of several species has brought resounding success to the state's major river systems including the Connecticut, Merrimack, and five coastal rivers. The coastal rivers—Cocheco, Exeter, Lamprey,

Oyster, and Taylor—have developed major runs of river herring over the ladders. By transporting spawning adults to Strafford's Bow Lake, which forms the headwaters of the Cocheco River, the number of fish returning to the ladder in Dover increased from 477 in 1984 to over seventy-two thousand in 1992.

The Merrimack River has had an increase in river herring from about a thousand in 1984 to nearly four hundred thousand by 1989 in the lower river, and with the completion of a fish ladder at the Amoskeag dam in Manchester in 1989 has allowed herring to reach all the way up to Hooksett. The Atlantic salmon run in the Merrimack increased from 65 in 1988 to 332 in 1991.

The Connecticut River along our western border has had substantial increases of Atlantic salmon, American shad, river herring, and sea lampreys over the two fish ladders south of New Hampshire allowing them once again to swim into our state's waters. Between 1979 and 1990 the American shad run increased from 300 to 27,900 at the Turners Falls Dam, which put these fish into New Hampshire's waters.

The trend for increased wildlife populations for both fish and wildlife will likely continue over the long term.

If you are a hunter, fisherman, or even a casual wildlife observer you have a greater likelihood of experiencing these wildlife species than your grandmother and great-grandmother did. The good old days are now!

*First published in 1992.*

# 22 Million Juvenile River Herring Seeking the Sea

US Fish and Wildlife Service biologist Joe McKeon, from the Laconia office, is predicting that upwards of twenty-two million juvenile river herring will be heading down the Suncook River into the Merrimack River. This slurry of two- to four-inch-long silvery fish must dodge great blue heron, gulls, otter, and some very hungry bass. They also must go over, around, or through the seven hydroelectric dams that are scattered the length of the two rivers. How many will successfully reach the sea at the mouth of the Merrimack depends greatly on the willingness of the dam operators.

Two years ago, perhaps millions were made into sushi for the hordes of gulls at the confluence of the Suncook River and the Merrimack River when low water conditions sent all the young fish into the turbines.

In 1997, the hydro operators were alerted to the downstream migration of up to fifteen million fish when the dam-boards at Crystal and Suncook Lakes in Barnstead and Northwood Lake were removed to lower the lake levels for the winter. The torrent of water sent the anxious juveniles into the waiting currents of the river for the quick ride out to sea. Most of the juvenile fish were allowed safe passage past the hydro sites thanks to the cooperation of the operators.

The river herring restoration effort was initiated about six years ago by Bill Ingham, biologist with the New Hampshire Fish and Game Department. I worked with Bill to transport several hundred adult alewives (river herring) from the seventy-two thousand that had ascended the Fish and Game fish ladder on the Cocheco River in Dover to Northwood Lake. Adult alewives migrate up the coastal rivers in the spring from the sea. There are so many that some can be transferred inland to the Merrimack River system in order to bolster the ongoing restoration efforts there. One adult female alewife may lay two hundred thousand eggs! With the assistance of the staff from the marine division, several hundred were released in Northwood Lake in Northwood. Early success was evident by the large schools of young alewives seen patrolling the lake shores by midsummer.

About four years ago the US Fish and Wildlife Service offered their assistance to help transport larger numbers of adults. What a difference they have made! In 1998, there were 7,845 adult alewives transferred to the tributaries of the Merrimack.

Fish passage for upstream migration has been constructed at the three lower dams, which allows fish to swim up to the dam in Hooksett. By transferring adults who spawn into the tributaries it is hoped that a greater number of fish will return to the Merrimack River and more quickly restore the ecology of the whole system. River herring can provide a significant forage base for the many predators living along the river and even the gluttonous striped bass that dominate the food chain where the river meets the sea. An additional twenty-two million herring can only mean good things for the Merrimack River system.

*As of 2024, the number of herrings continues to grow. They've been restocked in Lake Winnipesaukee, which has been more successful than any other attempt. This is where they should have been all along, because thousands of years ago, Native Americans fished thousands of herrings at the weirs in Lake Winnipesaukee.*

*First published in November 1998.*

# New Hampshire's Threatened and Endangered Wildlife Rebounding from the Mountains to the Sea

From the high peaks of the White Mountains where peregrine falcons soar, to the sands of Seabrook beaches where the piping plovers scamper, New Hampshire's threatened and endangered species continued to rebound in 1998.

Ten pairs of peregrine falcons nested along the cliffs of the White Mountains and to the north. Sixteen young were fledged from the seven successful nests.

The spectacular osprey was able to fledge twenty-five chicks from eleven successful nests despite the floods of June that dampened their nesting ability. Although the osprey stronghold is the area around Umbagog Lake in Errol, the Seacoast continues to have four nesting pair. The excitement continues to be the expansion of osprey into the central part of the state along the upper Merrimack River system.

The lone pair of nesting bald eagles continued with their success in the waters around Lake Umbagog in Errol as well. Unexpectedly, a pair of eagles attempted nesting in the southwestern part of the state. We have high expectations for next year's nesting season for this pair.

An experimental project begun in 1997 on Seavey Island at the Isles of Shoals sought to return nesting common terns that were displaced by gulls several decades ago. Last year this was met with limited success. But in 1998, forty-five nesting pair of common terns were enticed to nest on the island by coaxing them with a flock of wooden decoys and a CD of raucous calls of other common terns played so loudly that only a teenager could appreciate it. Nearly one hundred young terns were fledged on the island in 1998.

Most exciting of all was the successful fledging of twelve piping plover chicks from five nests along the Seabrook Beach. This robin-sized, pale-colored bird had not successfully nested in numbers along our coast since the 1950s. Plovers first appeared on the New Hampshire coast again in 1997 when only three chicks were successfully reared from five nests.

This year's success was due to a cooperative effort by the New Hampshire Audubon Society, New Hampshire Fish and Game Department, the US Fish and Wildlife Service, and community officials from the town of Seabrook. Beachgoers were extremely cooperative as well. Meetings were held in Seabrook and plans were hatched well before the nesting season in order to develop the strategy that enabled twelve chicks to fledge compared to only three last year. Few people, except for those directly related to the project, realize the effort needed to bring a species back from the brink of extinction. As the wildlife biologist for the coastal region, I was called upon at the first sighting and immediately gathered equipment to protect the first nest, and rushed to Seabrook to protect the nest. It has given me great satisfaction to know that my years of effort on this project have helped restore this federally endangered species.

*As of 2024, there are over three thousand nesting pairs of terns on the Isles of Shoals. In 2023, there were over twenty nesting pairs of piping plovers on New Hampshire beaches.*

*First published in September 1998.*

# Heavenly Habitat

The first astronauts were awestruck by the beauty and the fragility of the earth as it glided beneath them on each orbit. Well, since July 23, 1972, scientists have been able to measure that fragility thanks to the first Landsat satellite launched by NASA that summer day. Other Landsat satellites were launched regularly over the last three decades to continue the work of looking down at earth from space to monitor vegetative changes worldwide. The latest Landsat satellite, Landsat 7, was launched April 15, 1999, to continue one of the longest and most successful space-based programs.

Landsat satellites measure vegetative cover and numerous other Earth features by its Thematic Mapper (TM). Essentially the satellite takes continuous pictures of the Earth with specialized cameras that collect the light reflected off the vegetation. Millions of images have been taken. One of the best things is it takes images of the same location every sixteen days, so growing cycles, year-to-year changes, and even the impacts of development can be accurately measured over the long term. We humans are changing the face of the Earth. Landsat bears witness to those changes.

Beginning in 1991 the New Hampshire Fish and Game Department began funding a project to map land cover in northern New Hampshire. In 1990, legislation passed that required bear hunters to purchase a special bear hunting license. Incidentally this bill was proposed and shepherded through by the New Hampshire Bear Hunters Association. They had wanted a ten-dollar fee, which was scaled back to five dollars when it passed. But most importantly the funding was required to be spent on research and management of the state's bears exclusively. As the state's bear biologist at that time, I requested that we use bear money to begin mapping the bears' habitat. A large part of the bear license funds were first used to fund the Fish and Game's portion of the land cover mapping. These and other hunter fees were the backbone of a more recent statewide land cover mapping project that the department helped fund.

Fortunately, the NH GRANIT (New Hampshire Geographically Referenced Analysis and Information Transfer System) staff at the Complex

Systems Research Center at the University of New Hampshire was already working on a similar project for the EPA in southern New Hampshire in 1991. The ten thousand dollars in bear license fees was used to expand the mapping to the northern counties in order to access bear habitat. In the initial mapping of the northern regions, nine different land cover types or classes were generated using the TM imagery. These included: hardwood versus softwood, various wetland types, and other land uses such as agriculture, gravel pits, and alpine habitats.

In January 2002, the NH GRANIT staff at Complex System completed a statewide New Hampshire Land Cover Assessment. This most recent mapping effort used Landsat 5 and 7 satellite images collected between 1990 and 1999 to develop a statewide map of twenty-three types of land cover. This time the satellite imagery was able to distinguish types of trees such as beech/oak, spruce/fir, hemlock, and birch/aspen, and even in some cases how dense the forests were. Most importantly, thousands of ground-truthing sites were selected and staff was sent to check each one in order to improve the accuracy of the satellite maps. As a result, the current data is known to be 82 to 95 percent accurate, depending on the forest type. This was a vast improvement over the last mapping effort. The New Hampshire Fish and Game Department contributed eighty thousand dollars to this two hundred and fifty thousand dollar project. Several other state and federal agencies contributed to the latest mapping effort.

Now conservation groups, such as the Bear-Paw Regional Greenways, can use this important data to develop conservation plans to protect and preserve the most significant wildlife habitat. These land cover types will be used to analyze wildlife habitat and travel corridors needed to connect Bear Brook State Park with Pawtuckaway State Park as well as other significant habitat blocks in western Rockingham County and eastern Merrimack County. Thanks to this latest Landsat technology, New Hampshire is better prepared to meet the challenges of identifying and conserving vital wildlife for decades to come.

*As technology evolves, NOAA continues to launch satellites that more accurately access ground habitat.*

*First published in August 2002.*

# Silent Woods

Nearly a half century ago Rachel Carson's book *Silent Spring* awakened a giant, the American public, to the environmental disasters caused by DDT and other chemicals. The awakened giant was motivated to outlaw the use of this product in the United States. An environmental movement hatched after a brief incubation period in the early 1970s that included the first "Earth Day." A fledgling peregrine falcon restoration effort took off as well. These birds were vastly diminished in numbers across North America because DDT caused them to lay thin-shelled eggs that failed to hatch. New Hampshire is fortunately the residence of over a dozen pairs of nesting falcons because of the restoration efforts, the banning of the dangerous chemicals, and the release of falcons in New Hampshire.

Yet a more insidious disaster has infiltrated our wild lands across New Hampshire and much of North America. It is not poured from a barrel, is all around us, yet is unseen, and has caused an even more dramatic decline in bird numbers far greater than Rachel Carson could have imagined. It is called urban sprawl and its result, habitat fragmentation. The disaster is us. It is you and me, our sheer numbers. It is our houses, roads, golf courses, malls, parking lots, and whatever else we have deemed necessary for human civilization.

Our wanton ways of using this land we call New Hampshire has caused a significant decline in numbers of songbirds, called neotropical migrants, in the last two decades. Neotropical migrants are the colorful songbirds that live and breed here through the spring and summer, yet must migrate vast distances to Mexico, Central and South America, and the Caribbean in the winter. Less than 10 percent of the songbirds we love to hear and watch in the spring actually winter here. Most are neotropical migrants. For instance, blackpoll warblers migrate nonstop for twenty-three hundred miles over an eighty-six-hour commute from eastern North America.

Red-eyed vireos, scarlet tanagers, song sparrow, meadowlark, ovenbirds; the list goes on and on. Over twenty species have been recognized to be in decline locally.

Studies over the last two decades have laid much of the blame on the decline in loss of large tracts of land to development, those over five hundred acres. Indeed, urban parks protected decades ago have fallen silent to many of the wooded songbirds historically found in them. They have become "silent woods." For a variety of reasons, the fragmented woodlands have become sinkholes to songbirds, particularly ground-nesting birds like the ovenbird. Predators such as skunks, raccoons, and even our lovely housecats efficiently sweep these smaller woodlots clean during their foraging. Skunk and raccoon numbers can explode with little natural control thanks to the availability of our trash and pet foods. The smaller lots also increase the "edge effect" that actually encourages several species, such as jays, starlings, cowbirds, orioles, and crows, which also prey on the eggs and young of neotropical migrants.

At least in New Hampshire, most land conservation efforts are at the town or several-town level through the efforts of local land trusts. My experience has been that just a few key people in a town or adjoining town can really make a significant different in protecting large tracts of land that will maintain a diversity of species. Will your favorite woods be silent? Or will your grandchildren hear the melodious notes of a wood thrush? You can give song to our forests of the future through your town's conservation commission or local land trust.

*These woods were good medicine for us during the COVID-19 pandemic. First published in 1999.*

# Bear-Paw Greenways: Woodland Tracts for Tracks

The Bear-Paw Regional Greenways Project is a grassroots effort by citizens in a seven-town region of northwest Rockingham County and southeastern Merrimack County to string together ribbons of greenways connecting significant wildlife habitat in southeastern New Hampshire. *Why?* would be the typical question that is asked of the group. After all, aren't there animals nearly everywhere? "Why, just the other day a fisher cat dashed across the road in front of my car at dusk, just down the hill from my house."

The fact is New Hampshire has an abundance of wildlife not seen in nearly two centuries. Wildlife *is* everywhere. Over the last two decades the black bear population has grown from fifteen hundred to nearly five thousand and has spread southward in the state to some of the towns in Rockingham County. Moose literally are everywhere, lumbering from fewer than fifty several decades ago to nearly nine thousand now, also spreading north to south, and now have taken up residency in nearly every town in the state. Turkeys have gobbled up the available habitat for them as well across the state and now number twenty-four thousand. Fisher, foxes, coyotes, and a whole host of birds and mammals now call this corner of the state home. Trouble is, so do more and more humans at an ever-quickening pace.

People are shredding the wildlife habitat and fragmenting the large blocks of habitat that these critters once called home at an ever-increasing rate. It has been estimated that fifteen thousand to twenty thousand acres of habitat are lost each year. The majority of this loss has been in the southeastern side of the state.

Fortunately, there are permanently protected, large parcels of land in the form of state parks in this part of the state. Bear Brook State Park, encompassing 10,083 acres, is the largest permanently protected tract of land in the area and Pawtuckaway State Park, with 5,536, is the second biggest protected tract. There are others in the county as well, including Northwood Meadows State Park with 664 acres.

Although each of the larger tracts certainly can harbor a host of species for the foreseeable future, the rapid rate of human development is threatening to make these areas "islands of wildlife habitat in a sea of human development."

The Bear-Paw Regional Greenway Project seeks to sew these large blocks of protected habitat with ribbons of greenways to allow for natural movements of wildlife between these and other large blocks of habitat. There are five basic needs that the planned greenway wildlife corridors will provide: 1) Wide-ranging animals can travel, migrate and meet mates. 2) Plants can successfully propagate. 3) Genetic interchange can occur between the isolated populations. 4) Populations of wildlife can move between blocks in response to environmental changes or natural disasters. 5) Individuals can recolonize habitats from which wildlife populations have been locally extirpated.

Studies have shown, and it makes sense, that a wildlife corridor is only as strong as its weakest link. Your house and your neighbor's *is the weakest link*. However, chances are you will not be voted off no matter how the votes are counted. So, the next best thing is to secure enough habitats and permanently protect and manage it so that the wild things we know make New Hampshire so special will always survive. Maybe things such as a bobcat under a bird feeder in the winter, or a moose chest deep in a beaver pond eating lilies on a hot summer's day, or a coyote howling on a crisp February moonlit night. These are your neighbors now; wouldn't it be nice to share our state with them for another hundred years! Bear-Paw seeks to give them a home and ways to roam for the foreseeable future.

*The Bear-Paw Greenway Project is ongoing, working to protect more land every year.*

*First published in July 2001.*

# It's Duck Season in New Hampshire

It's duck season in New Hampshire right now. Some of you may be thinking, "Hey, we have no spring duck season?" Maybe I should say it is the season for ducks in the Granite State.

While I have loved observing ducks nearly my whole life, I should actually fess up to falling in love with wood ducks a long, long time ago and in fact have been having an affair with them for over fifty years. From the gleaming iridescent head of the drakes to the high-pitched squeal of a hen taking to flight, there is just something about wood ducks that stirs my soul.

Let me tell you how I first fell in love with wood ducks. This will take me back over five decades now to about 1963. Yes see, my parents had moved the family from far northern Maine to Londonderry in 1962 and to my wonder I discovered a marsh not a half-mile hike through the woods and fields behind my rural home. There I found Little Cohas Marsh. This is an over two-hundred-acre marsh created by a dam New Hampshire Fish and Game constructed in the late 1950s. I have always loved marshes too, or as my mom was quick to call them, "swamps." A term I still favor for an afternoon of adventure.

It was at Little Cohas that I first spotted some strange wooden boxes attached to the trees in the marsh as I hiked to the marsh over that winter (1963). I began my research on these boxes and soon learned they were duck nesting boxes. That next year I took a hunter safety course given by the local Londonderry Fish and Game Club. By the end of the course, I was amazed that the men of the club liked wildlife and they liked ducks a lot too. So, it was not that hard to convince them to fund my efforts to build another eight or ten boxes that winter (1964). I added a few more boxes to Little Cohas Marsh and a club member helped me put up the rest in other marshes in town.

By that next winter I began "checking" my duck boxes to see if they were used or not. And I faithfully checked them each winter on through high school. I can tell you right off that it is best to check them in winter. I remember taking a small boat out to them in early summer one year only to find a hornet's nest in one that had me stung up a bit. Another time I stuck my arm in the hole only to have a mouse run up my sleeve and into the back of my shirt as I frantically untucked it to let the mouse out. Yes, there always are some surprises to be found besides ducks. By college at UNH I can't say that I checked them as regularly but years later I learned that in fact the Fish and Game waterfowl biologist was taking care of them. Although as part of a college project I actually checked a number of duck boxes the professor had put up at some local marshes near campus. I found ways to stay in love with wood ducks even while away from home.

Fast forward a bit to the mid-1980s. Here you will find me actually working for the New Hampshire Fish and Game Department as a wildlife biologist. By the late 1980s the department had built regional offices. I chose to not move and I became the Region 3 wildlife biologist out of the Durham office. Guess what one of my assignments became? Wouldn't you know it. It was to check the wood duck boxes each winter in Little Cohas Marsh along with others across the southeast part of the state. Talk about coming full circle. Here I was going to be paid to do what I wanted to do most as an early teenager. So I checked the boxes each winter at Little Cohas right up to my retirement in 2007. And I found a few more surprises. Like on one Groundhog Day in the 1990s when I opened a box at Little Cohas only to find a hibernating meadow jumping mouse. I was kind and just tucked the little fellow back into the box. I found a honeybee hive one year as well.

I should also mention that not only are the boxes checked during the winter but each May for a number of years I also checked a few to capture female wood ducks for banding. So I would canoe up to each box and block the hole and carefully take off the side panel to capture the female for banding. This was a pretty efficient way of banding female wood ducks. I remember one hen that I banded in a box in Bear Brook State Park returned to the same box several years in a row. I should also mention that hooded mergansers are the other cavity nesters that use these boxes.

I can guarantee you that the duck boxes at Little Cohas Marsh are still being checked each winter. In fact, I recently called Fish and Game waterfowl biologist Jessica Carloni to learn the results of recent winter checks. Below is her summary:

At fifty marshes during the reporting period, two hundred and twenty functional wood duck nesting boxes were inspected for use by nesting wood ducks and hooded mergansers. Evidence of use by waterfowl was observed in 136 (55.3 percent) of the functional boxes that were inspected. This was higher than last year but lower than the past few years: 2013 (52.6 percent), 2012 (63.9 percent), 2011 (69.0 percent), 2010 (68.6 percent), and 2009 (72.7 percent).

Wood ducks accounted for 21.3 percent of the observed use, which compares to 21.7 percent in 2013, 29.8 percent in 2012, 25.3 percent in 2011, 23.5 percent in 2010, and 29.3 percent in 2009.

Hooded mergansers occupied 57.4 percent of the nest boxes used by waterfowl compared to 62.3 percent in 2013, 63.9 percent in 2012, 67.8 in 2011, 70.6 percent in 2010, and 63.9 percent in 2009.

Hooded merganser and wood duck eggshells were observed in the same box on twenty-nine occasions.

Waterfowl were considered to have been successful in hatching young in 75.9 percent of the nests that were initiated, which was similar to the past five years: 62.9 percent in 2013, 72.3 percent in 2012, 71.3 percent in 2011, 76 percent in 2010, and 75.9 percent in 2009.

*First published in March 2016.*

# We Have a Lot to Be Thankful For in New Hampshire

Just a couple of days ago as I sat here at my computer, a glance out my window that rainy day showed a partridge picking the fruit off the high bush cranberry shrub I planted years ago. My crabapple trees are also loaded with fruit so I'm hoping to see more of my local partridge.

Heading to Concord the other day I had to pull over before entering the state road to just soak the sight in. For across the highway in the now-cut cornfield was a flock of forty turkeys flanked by more than a dozen geese. In fact, I see turkeys practically every day; indeed I spotted a flock in the field across from my house just yesterday.

I have been saying for years that we are living in the "Golden Years of Wildlife." And in fact, I had a slide program to that regard I gave to the public the last decade or more of my Fish and Game career.

While climate change may be whittling away at our moose population—now down to about four thousand from seventy-five hundred a dozen years ago in 1999—most species have held their numbers or have grown. Even moose numbers are way higher than just four or five decades ago. There was believed to be fewer than fifty moose fifty years ago. Though numbers are down they still occur over a wider range and in greater numbers than they had for over two hundred years.

Our milder winters as a result of climate change has meant an ever increasing number of deer. Now pegged by Fish and Game at over one hundred thousand, deer were very scarce over much of the state well into the early 1900s. Although the state's record kill came in the mid-1960s it was the result of a two-month split season with a month to the north of the Lakes Region in November, then another month south of there in December. So you could hunt for any sex deer for two months. I remember by the early 1980s, with a deer kill of just over three thousand, that Fish and Game was figuring a population of about forty thousand deer. So our deer numbers have more than doubled in the last forty years.

There were *no* turkeys when I grew up in southern New Hampshire in the sixties and into the seventies. Finally, the 1975 transfer of turkeys from western New York took hold and off they went. Hard to believe that with a population of just twenty-five turkeys in 1975 we would have such abundance now. I think the most recent estimate is closer to forty thousand wild turkeys out and about. Seems like they are everywhere.

Geese? There were no geese when I grew up in Londonderry, save for a few high-flying V's that grabbed your attention as they etched the sky and cast down that wonderful music to a hunter's ear. Well, you could drive over to Great Bay during the late duck season and gawk at the thousand or so geese rafted up in the middle of the Bay. They seemed no more accessible than the highflyers. I spent hours with my best friend Rick laying on a cold ice slab hoping one would fly our way.

Now thanks to the "resident" geese that moved into the state in the 1990s we are practically overrun with geese. I think the estimate is something like forty thousand resident geese now. We have a more liberal season and bag limit to take geese than ever before.

When I became the state's first bear biologist in the late 1970s I started collecting bear teeth from successful bear hunters. By the early 1980s I had collected enough information to estimate we had about a thousand to twelve hundred bears in the state. But the bad news was there were few older females. It looked like bear numbers were in a decline. Thankfully with data in hand the legislators turned over the management of bear to Fish and Game in 1985. We cut the season from three months to one and closed the southern seven counties to bear hunting altogether. That is what it took to turn the decline around and bring bear numbers to closer to five thousand today. Bears now occur over a wider range and in greater numbers than in over two hundred years.

Even animals like beaver and raccoons were not all that plentiful. In fact, there were season limits on how many a trapper could take of beaver, and hunters could take of raccoons, into the 1960s.

Beaver were essentially gone from New Hampshire by 1900. There was believed to be one colony left on the Canadian border when they were finally protected in 1901. Beaver numbers slowly grew and Fish and Game staff actually moved some south to Holderness in the 1930s. Beaver really were not all that common until the mid-1950s into the early 1960s as I recall. Hence the seasonal bag limits for trappers. Now beavers are living in likely all accessible habitat. Trappers take several thousand a year.

And thanks to our wonderful beavers' engineering practices, we have lots of wetland habitat for other species. Mink and river otter are now common across the state.

And duck numbers have grown as well, as the beavers made more and more habitat for them. Mallards were a rare duck when I first started duck hunting in 1966. The last I knew there was something like an estimated ten thousand nesting pairs of mallards in the state.

When I started working at Fish and Game in 1976 there were no bald eagles or falcons nesting in the state. There were a couple pairs of ospreys up in Errol as I recall. In the 1990s, I helped New Hampshire Audubon count nesting common terns around Great Bay. The sum total most years was eight.

Now we have something like forty pairs of nesting eagles, fifty osprey nests, twenty-two pairs of falcons, and over twenty-five hundred pairs of terns where a population was restored at the Isles of Shoals. And how handy have the terns been some days I have been striper fishing on the bay. I remember one day I simply drove from one group of diving terns to the next for several hours catching stripers at each stop.

Yes, there is so much to be thankful for.

*First published in November 2015.*

# Endangered and Threatened Birds Making a Comeback in New Hampshire!

Several of New Hampshire's threatened and endangered bird species continued to make a recovery during the 1999 nesting season. Recovery for any species is often a long and tedious process fraught with challenges at every turn. The balance of nature once upset by environmental causes, such as the decline in several bird species due to DDT, can take decades to restore. Such was the case for several of New Hampshire's most elegant bird species like the bald eagle and the peregrine falcon.

Thankfully, the causes for these disastrous declines were discovered and to a large degree have been removed from the environment. A new respect for the environment, which really blossomed in the early 1970s, has come to fruition nearly three decades later. This generation is a witness to the power of nature, with man's help and understanding, to once again grace the skies of our state with magnificent birds like the eagles and falcons.

The 1999 breeding season with long periods of dry favorable weather allowed for record-setting successes for several species. The peregrine falcons hatched a record twenty-five young from eight successful nests. A total of twelve pairs of falcons were observed frolicking among the state's highest peaks. Bald eagles were active at three nest sites scattered across the state. The pair that has nested for a decade in Errol was successful again. Two new pairs of eagles set up housekeeping elsewhere in the state and hopes are high for their success next year. Eagles were absent from the state for over fifty years. Welcome back, glorious eagles!

The beautiful osprey, a bird who dives headlong into the water to catch its prey of fish, has made a remarkable return as well. This year there were thirty nesting pair. The newest pair adopted an artificial nest placed high in a tree on the shore of Lake Massabesic in Manchester, by the Audubon Society.

There were a couple of record-setting successes along the coast as well, due in large part to the tireless efforts of staff from the Fish and Game Department, Audubon, the Fish and Wildlife Service, and local volunteers. The robin-sized piping plover, a federally endangered bird species, had a record-setting five nests along the sandy beaches of Hampton and Seabrook. A whopping sixteen young were produced! Local beachgoers were kept informed of the birds' presence so the thumbnail-sized chicks could survive unharmed by human activity. "It takes a village" to successfully raise five families of plovers! Out at the Isles of Shoals a record-setting 138 common terns nested. Over a hundred chicks have been banded so far with more to come. This is nearly triple the number of nests from just three years ago.

Several of New Hampshire's threatened and endangered birds are making a strong recovery thanks to a healthy environment and a little help from humans. This is a good start but much work remains to be done. Hopes are as high as the soaring birds that next year will bring more record-breaking successes.

*As of 2023:*
*Over three thousand pairs of terns nest on the Isles of Shoals.*
*Over one hundred pairs of bald eagles nest in the state of New Hampshire.*
*Nearly two hundred pairs of ospreys nesting statewide.*
*Twenty-nine pairs of peregrine falcons and*
*Fifteen pairs of piping plovers nesting on the coastal beaches.*

*First published in 2000.*

# The Eye in the Northern New Hampshire Sky

In late November and early December 1998, moose in northern New Hampshire were counted using the latest in technology: an infrared camera mounted on the wing of an airplane. For the first time, wildlife biologists were able to observe moose across a vast section by using a method to accurately "see" moose from the air.

The Fish and Game Department contracted Airscan, a Florida company that utilizes a gyroscopically controlled infrared camera mounted under the wing of a twin-engine airplane that records the images on video.

While it would be impossible to actually count every moose in the state, or even just those in the North Country, biologists devised a plan to sample likely moose habitat across the North Country in order to provide a statistically valid estimate. Biologists selected forty plots of about six square miles each for the plane to scan for the presence of moose. Only likely moose habitat was scanned and large water bodies; developed areas and agricultural land was skipped. About 258 square miles were checked by the camera.

The infrared camera actually senses the heat given off by the moose's body. They actually glow white, especially in the dark. The video I saw had the unmistakable images of white moose through the dark trees. Even the antlers could be detected on the bulls. This technology is amazing!

Late November was chosen to start the count as the moose would be more visible under the leafless trees and the department wanted to get a count as soon after the deer hunter mail survey as possible. Each year successful deer hunters from the previous fall are sent a questionnaire that asks them to keep a diary during the first ten days of the current deer season, noting how long they hunted and how many moose they saw. This survey has been conducted for several years and has been used to gauge the relative densities of moose across the state. By comparing the number of moose seen per one hundred hours of hunting, our moose biologist has been able to determine where moose were more abundant. However, this data could not be used to calculate actual moose numbers. This is why an aerial survey was needed.

The aerial survey not only gives the department its first scientific estimate, but it can also be used to make the hunter survey more useful. The airplane actually circled each of the forty plots and scanned them to determine how many moose were in each circle. As you would expect, some sites held few or no moose, but one plot had over one hundred moose! The plane was able to fly and survey on twelve days during its nearly monthlong stay at Berlin Airport. Bad weather precluded flights on a number of days. Hey, it's the Great North Woods. No one said it would be easy, but safety for the air crew came first.

The most accurate survey, and therefore estimate of numbers of moose, was made for Coos County where the actual flights took place. The moose population was pegged at forty-one hundred. By combining the northern count with the hunter surveys from elsewhere in the state, biologists were able to estimate moose numbers across the state. The statewide population was estimated at ninety-six hundred. Moose densities range from a high of over 3 per square mile in the north to about 1.29 in the White Mountains Region, 0.95 in the Central Region, 0.89 in the southwest, and 0.80 in the Southeast Region.

In 1998, the information needed to manage the state's moose population improved dramatically by the use of the aerial infrared survey. Further refinements in the technology over the next three to four years will likely generate more accurate estimates. These numbers are subject to change! After all, moose do moose things. There is so much to learn about the giants of the New Hampshire forests.

*New Hampshire Fish and Game is no longer doing aerial surveys for moose. They have not done so for a number of years.*

*First published in February 1999.*

# RX-deer: Taking the Pulse of New Hampshire's Deer Herd

Each fall New Hampshire Fish and Game wildlife biologists and technicians scatter across the state and set up "deer biological check stations" at over twenty sites. One of the sites is even operated by a biologist from the US Forest Service in order to collect as much information as possible, particularly during the either-sex days of both the muzzleloader and regular firearms seasons.

The biological staff assess the condition of the state's deer herd each year by examining thirteen hundred to fifteen hundred deer. Biological check stations are operated in conjunction with already-existing mandatory deer registration stations that are usually located at sport shops or local convenience stores. Sites are selected based on the number of deer that are usually registered at the station, and its location in one of the state's Wildlife Management Units (WMUs). Most of the sites have had staff at them for nearly two decades, often the same person, so a great working relationship has developed over the years with the station owners. The folks who own and operate these deer registration stations have served the sportsmen and the department well for many decades. They continue to play a major role in collecting most of the information on the deer harvest each year including date of kill, sex and relative age, fawn versus adult, town of kill, and lots of information about the successful hunter. The biologists supplement this information with biological data.

The biologist's role is to collect data on the weight of the deer and examine their teeth in order to age each deer. Most deer can be aged to the correct year by comparing the development of their teeth up to age one and a half. Then tooth wear is used to determine their age. Deer gradually wear out their teeth over the next few years. There are fine detectable differences in the amount of wear each year that are visible to the trained and experienced eye. In New Hampshire very few deer get to be eight or nine years old and the few that do have worn their teeth down to the gum line or below.

Although a variety of data is collected, one of the most significant gauges of the health of the herd is the antler beam diameter of yearling bucks. Good antler development in these deer signifies a healthy population. In 2000, the average was 17.1 mm statewide, slightly below the five-year average of 17.4 mm. About 47 percent of the annual buck kills are yearling bucks. Yearling does accounted for 34 percent of the does in 2000 compared to a five-year average of 26 percent.

Other interesting bits of information include: About 10 percent of the yearling does were milking, indicating that they had bred as fawns. About 54 percent of the adult does were milking. About 40 percent of the yearling bucks had spikes only, so most of these young bucks are already sporting forks or more. Yearling buck weights averaged 114.1 pounds compared to 115.9 in 1999.

Thanks to license fees collected from hunters and the Sportfish and Wildlife Restoration funds obtained from a special federal tax on arms, ammunition, and archery equipment, over eighty-four thousand dollars is spent each year to manage the state's deer herd. It costs about five thousand dollars a year to collect the very important biological data that is needed to fine-tune the health of our deer population.

When it comes time to check the condition of New Hampshire's deer herd you can be sure that the "doctors are always in."

*This biological data collection is ongoing.*
*First published in December 2001.*

# Wild Goose Tails

Since 1991, New Hampshire Fish and Game wildlife biologists, technicians, and volunteers have flocked together for the annual "wild goose roundup" that occurs the last week of June. Someday there will probably be a country-western song written about this event. It truly is a week of heat, and dust, and feathers, and sweat, and goose crap, and blood (ours and theirs). I can well imagine the possibilities of titles to the song.

The "goose crew" led by waterfowl biologist Ed Robinson meets each morning to formulate the plan for the day's event. For about a ten-day period at the end of June the adult Canada geese molt most of their flight feathers and are grounded. The goal of the goose crew is to sweep across the state, particularly the southern counties, and round up and tag as many geese during this annual molt as possible. A minimum of six to eight goose boys or girls are generally needed to surround the flock and successfully herd them into a hastily constructed pen, usually situated at the end of some pond. Geese seek small ponds, some rivers and lakes with an ample supply of green grass to graze on nearby, and the water for escaping predators during the flightless period. And boy do they eat grass. But more significant is the amount of "material" exiting from the southern end of a north-facing goose. A goose can deposit one and a half to two pounds of goose crap a day. It seems like at least a pound of it can be found under your fingernails after a day at the roundup!

After all we are not just rounding them up but each one needs to be handled in order to determine its sex and age. This process requires a person to swing the goose upside-down and pin it to the ground between your knees as you very carefully and not so delicately open its butt by squeezing it. Hence the pound of goose "doo" under the nails. Geese have also obviously attended gunnery school, as they are quite accurate at shooting the stuff some distance as well, usually on the cleanest individual or unstained piece

*New Hampshire Fish and Game goose banding team.*

of cloth or arm within range. A light-colored to pinkish anal vent means that the goose is only a year old. Adult geese have a dark gray to nearly black-colored anal vent. Males will have a corkscrew appendage evident.

The geese also receive a nice shiny leg band with a unique number inscribed on it as well as instructions to call an 800 number to report it. Since 1991, 7,021 geese have been banded in New Hampshire. In 1991, 1992, and 1993 some geese were also fitted with white neck collars, also with unique lettering or numbers so the geese could be identified from a distance using binoculars or a spotting scope. It was the results of these first few years of study that the "resident" goose population was identified in the state. These geese spend nearly their entire year, in fact their entire lives, in New Hampshire or not far away. A mild winter, as we have had for most of the last decade, finds these birds spending the winter at small open water spots at office complexes or condo developments just above the Massachusetts border. A winter with prolonged freezing may send the birds south into Massachusetts and occasionally further, but the first warm spell also spells their return. Although the neck collaring ended in 1993, a few birds still show up with collars on them.

This year in 2001 the goose crew was able to capture 762 geese. Monday, June 25, was the kickoff to the roundup. By 9 a.m. the temperatures were already in the eighty-degree range as 114 geese were persuaded to enter a pen in Rochester. Whenever possible the pen is placed in the shade to keep the geese as cool and stress-free as can be expected. A number of these geese had seen us before, as a cheering section of hisses could be heard from the amassed crowd of geese. Some of these guys were biters too. Geese have a ferocious bite that will leave a lasting welt and black-and-blue mark on wherever they have attached themselves. Most of the goose crew wears long sleeved heavy shirts or overalls despite the heat, to minimize the bites and scratches from their powerful feet, well-equipped with sharp nails to shred any exposed skin.

During the 2001 roundup, geese were captured at nineteen sites, including one in Northumberland. There were 173 recaptures and 589 new birds were banded. There were thirty-four goose broods identified with an average of 5.3 goslings per brood. Four geese with neck collars were recaptured this year including two in Atkinson and one each in Jaffrey and Northumberland. All four had been collared in 1993, the last year it was done.

New Hampshire's resident goose population has continued to explode during the last decade and now numbers over thirty thousand birds. While the majority of the birds are located in the southern tier of counties, the fact is resident geese now live in every county. Beginning in 1996 a special resident goose-hunting season has been held during the month of September to try to reduce the dramatic buildup of resident geese. Many of these geese live in urban or even city complexes so have few predators including hunters because of their urban hideouts. However, hunters have taken an average of 1,790 geese a year over the last five-year period during the special season. In 2000, hunters were estimated to have harvested 2,010 geese during the special season. Additionally, another 4,534 geese were taken during the traditional fall season, making the total of 6,544 a new state record harvest. Most of these geese were from the resident flocks as well. A few migratory geese are taken primarily on the Great Bay waters in late season. In fact, the last three years hunters have set new goose harvest records in the state, with 3,691 taken in 1998 and 6,072 taken in 1999. There is little doubt that the record could be broken again in 2001.

New Hampshire's resident goose season is slated to run from September 4 to September 25 statewide with a daily bag limit of three. Don't forget to purchase both a federal and state waterfowl stamp to participate in the hunt. Scout early and be sure to get landowner permission well in advance of the season. Choice hunting areas are sewn up early by hunters who have developed a good relationship with the landowner. Volunteers are always welcome so have a great fall hunting and think about joining the goose crew next June.

*As of 2024, there are over forty thousand resident geese in New Hampshire. First published in 2001.*

# Just Ducky in New Hampshire

New Hampshire's waterfowl populations are thriving thanks to nearly two decades of conservation efforts by a host of groups. New Hampshire's hunters have supported the conservation efforts through the purchase of licenses, federal and state duck stamp sales, and funds from the Sportfish and Wildlife Program (a federal tax on the purchase of firearms and ammunitions).

In the mid-1980s, waterfowl populations across North America were in a downward spiral. Although several years of drought conditions in parts of North America played a significant part in the downward trend, a plan was developed to reverse the trend and the goal was to restore waterfowl numbers to numbers common twenty years before. In 1986, a North American Waterfowl Plan (NAWP) was adopted to begin the restoration efforts.

In New Hampshire, the plan was focused on the restoration of the black duck, whose numbers had plummeted since the mid-1960s. Great Bay, a coastal estuary whose watershed includes nearly 25 percent of the state's land area, was chosen as the "poster child" for the plan. Great Bay has long been known for its role as the wintering grounds for thousands of waterfowl but most importantly, for the tremendous number of black ducks who cover the mud flats in the ebb of the tide. Over 80 percent of the wintering black ducks use the bay. Additionally, twenty species of waterfowl also winter on the bay, including a huge raft of migratory Canada geese, or stop there during migration. The large mud flats exposed during low tide provide feeding places for another twenty-seven species of shore birds and thirteen species of wading birds. Protecting and preserving Great Bay and its tributaries including buffers of upland habitat has been a priority for New Hampshire's plan.

Early successes were small. The state's four-dollar duck stamp, initiated in 1984, could only provide limited funding, even when matched with funds from the Sportfish and Wildlife Program by waterfowl biologist Ed Robinson. However, the momentum Ed started has grown, as has the amount of the bay under protection. The Great Bay National Wildlife Refuge set aside two thousand acres of shore on the north side of the bay when the former Pease Air Force Base was deeded to the US Fish and Wildlife Service in the early 1990s.

The pace picked up even faster after the passage of the North American Wetlands Conservation Act (NAWCA) in 1989. Thanks to the Waterfowl Plan, Great Bay has been a high priority for funding from this new act. Because of the quality of the waterfowl habitat of the bay, and the degree of threat by development during the booming economy, significant funding has been made available to protect the bay. In 1996, the first grant led to the protection of 535 acres, followed by 678 acres in 1998. Four grants have been funded so far, totaling $2,455,000, and leading to the protection of 3,815 acres.

Grant monies have been matched by seventeen other organizations, including The Nature Conservancy, Society for the Protection of New Hampshire Forests, Ducks Unlimited INC., New Hampshire Audubon Society, and Lyme Timber Company, totaling an additional $4,294,337. In fact a partnership was formed by a number of organizations in 1994 called the Great Bay Resources Protection Partnership to further the goal of protecting the bay. Through the partnership's urging, Senator Judd Gregg secured an additional seven million dollars in recent years to match funding acquired by the partnership. Much work remains to be done as the partnership has identified 14,200 acres as valuable habitat that deserves protection.

While the original focus was on Great Bay, the North Country had a gem of a habitat that needed protection too. Umbagog Lake, straddling the New Hampshire and Maine border, was singled out for protection under the NAWP. This area, too, has since been designated as a National Wildlife Refuge with permanent protection provided to thousands of acres of waterfowl habitat.

For 2000, the plan has been revised to increase the scope of the area around the bay and a new area, the Connecticut River Corridor, has been added to the plan.

The success of the NAWP has been resounding. Not just measured in acres of habitat protected but by actual numbers of ducks and geese present. Populations of waterfowl across North America have been restored. Some are at record levels! In New Hampshire, the decline in the black duck population has been stopped, although the numbers have yet to rebound to the former highs. Geese, well, geese just seem to be everywhere. Thanks to the conservation effort, mallard numbers are at near record numbers.

In 1999, the 3,900 hunters who purchased a state duck stamp took an estimated 21,600 ducks including 7,400 mallards, 3,200 wood ducks, and 1,900 black ducks. Teal, scaup, and sea ducks make up the majority of the other ducks taken. Although duck hunter numbers are down substantially from twenty-five years ago, their success at hunting is higher than ever thanks to the vision of the waterfowl plan.

*First published in November 2000.*

# 2022 Was a Banner Year for New Hampshire Hunters

What a year for New Hampshire's sportswomen and men who are Granite State hunters.

The year started off in May with a record spring turkey season take of 5,723. 2022 marks the fourth year in a row with a spring season take of over five thousand birds, a number not counted before 2019. Preliminary fall season numbers are 231 for archery hunters and another 460 taken during the fall shotguns season. Turkeys are abundant across the state with numbers never witnessed before.

New Hampshire black bear season, starting off September 1, came very close to setting the all-time record as well with 1,128 bears taken, just shy of a record of 1,183, taken just two years ago in 2020. This was the third year in the last five that more than a thousand bears were taken per year. Bear numbers are historically high across the state, estimated at over sixty-five hundred bears.

Deer number are up as well across the state. In fact, based on the current ten-year Fish and Game Deer Management Plan, the population numbers are above goal in fourteen of the twenty Wildlife Management Units. This has resulted in more liberal seasons in parts of the state resulting in ever-increasing deer takes. This set this year's deer season for a near-record take of 13,987, based on the preliminary tally from the results of conservation officers calling the deer registration stations on December 16, the day after the archery season closed. While not the final number, it is the third highest on record, and is close to the final results. The previous record deer kill was in 1967 when 14,204 deer were taken. But back then, female deer made up 49 percent of the kill versus last year when only 30 percent of the 12,551 deer killed were does. The annual buck kill has been climbing steadily since 1983. In the last forty-plus years the deer population has grown from an estimated number in the early 1980s of a low forty-something thousand deer to well over one hundred thousand today. This too is a historically high number of deer for New Hampshire.

Sadly, as New Hampshire continues to warm, our moose numbers continue to deteriorate. Only thirty-nine moose hunting permits were issued by lottery in 2022, resulting in only twenty-six moose taken, twenty-two bulls and four cows, for a success rate of only 62 percent. That is the lowest success rate since the season first opened in 1988. No moose were taken in five WMUs including D1, D2, E3, F, and M.

These really are "The Good Old Days" for New Hampshire hunters. Modern wildlife management by staff at the New Hampshire Fish and Game Department, with the help of others, like New Hampshire Audubon, over the last half century has brought about an abundance of game and other wildlife unheard of by previous generations. Other records continue to roll in for 2022. This includes bald eagles (ninety-two pairs), peregrine falcons (over twenty pairs), ospreys (over one hundred and eighty pairs), and out at the Isles of Shoals, terns (over three thousand pairs). Wow. You, yes, you have a greater chance of spotting any one of the above-mentioned wildlife than your grandmother or her mother ever did!

*First published in 2023.*

# About the Author

Eric Orff is a certified wildlife biologist with over fifty years of experience in New Hampshire. He graduated from the University of New Hampshire in 1972 with a bachelor of science degree in wildlife management. He worked at the New Hampshire Fish and Game Department from 1976 until his retirement in June 2007. From 1988 until retirement, he also served as the regional wildlife biologist for Region 3 in the Durham office. His specialty for the department was capturing wildlife. He captured and tagged numerous bear, moose, deer, and thousands of ducks and geese for the department.

Eric has received numerous awards throughout his career. In 2007 Eric received a certificate of merit from the State of New Hampshire Department of Resources and Economic Development, in recognition of thirty years of service with the Division of Forest and Lands and their Prescribed Burn Program. Eric was appointed by Governor John Lynch to the New Hampshire Fish and Game Commission in 2008, for a five-year term. In 2013 he received the Bell Ringers Award from the New Hampshire Fish and Game Department's Non-Game Division at their twenty-fifth anniversary celebration. In 2014 he received the prestigious Tudor Richards Award during the centennial celebration of New Hampshire Audubon.

In 2007, upon retirement, he also received awards from the Northeast Chapter of the Wildlife Society, the Nashua Fish and Game Club, and the Northeast Fur Resources Technical Committee. He was awarded lifetime memberships to the Londonderry Fish and Game Club and the New Hampshire Trappers Association.

Eric has spent decades as a popular writer, speaker, and now social media personality with his "What's Wild in NH" posts. Since 2004, he has posted frequent blog entries called NH Nature Notes, through his website, NHFishandWildlife.com. He frequently releases Facebook live videos on his own page, as well as on the New Hampshire Wildlife Federation's Facebook page.